ADVANCES IN DIVERSE APPLICATIONS OF POLYMER COMPOSITES

Synthesis, Application, and Characterization

ADVANCES IN DIVERSE APPLICATIONS OF POLYMER COMPOSITES

Synthesis, Application, and Characterization

Edited by

Suji Mary Zachariah

Yang Weimin, PhD

Maciej Jaroszewski, PhD

Sabu Thomas, PhD

AAP | APPLE ACADEMIC PRESS

First edition published 2023

Apple Academic Press Inc.
1265 Goldenrod Circle, NE,
Palm Bay, FL 32905 USA
4164 Lakeshore Road, Burlington,
ON, L7L 1A4 Canada

CRC Press
6000 Broken Sound Parkway NW,
Suite 300, Boca Raton, FL 33487-2742 USA
2 Park Square, Milton Park,
Abingdon, Oxon, OX14 4RN UK

Library and Archives Canada Cataloguing in Publication

Title: Advances in diverse applications of polymer composites : synthesis, application, and characterization / edited by Suji Mary Zachariah, PhD, Yang Weimin, PhD, Maciej Jaroszewski, PhD, Sabu Thomas, PhD.

Names: Zachariah, Suji Mary, editor. | Yang, Weimin, MD, editor. | Jaroszewski, Maciej, editor. | Thomas, Sabu, editor.

Description: First edition. | Includes bibliographical references and index.

Identifiers: Canadiana (print) 20220449503 | Canadiana (ebook) 20220449554 | ISBN 9781774910962 (hardcover) | ISBN 9781774910979 (softcover) | ISBN 9781003300526 (ebook)

Subjects: LCSH: Polymeric composites. | LCSH: Polymers.

Classification: LCC TA418.9.C6 A385 2023 | DDC 620.1/92—dc23

Library of Congress Cataloging-in-Publication Data

Names: Zachariah, Suji Mary, editor. | Weimin, Yang, editor. | Jaroszewski, Maciej Wladyslaw, 1966- editor. | Thomas, Sabu, editor.

Title: Advances in diverse applications of polymer composites : synthesis, application, and characterization / edited by Suji Mary Zachariah, PhD, Yang Weimin, PhD, Maciej Jaroszewski, PhD, Sabu Thomas, PhD.

Description: First edition. | Palm Bay, FL, USA : Apple Academic Press, 2023. | Includes bibliographical references and index. | Summary: "His new volume focuses on polymers, their characterization, and their various applications. These include drug delivery applications, electromagnetic shielding, ferroelectric applications, and many more. The book covers synthesis, characterization, and property studies of some of these polymers including their morphology, structure, and dynamics. It also introduces the most recent innovations and applications of polymers, fillers, and their composites in the electronics, biomedical, pharmaceutical, and engineering industries. Topics also include ferroelectric ceramics and the numerous polymers used for radiation shielding applications. The bottleneck in the development of novel technologies is often defined by the limitations of the available materials. Polymer science, for instance, has lately made large steps towards innovative materials and processes. Polymeric materials are ideal examples of structural and functional materials because of their versatility in practically all areas of modern life and technology. Some of their outstanding properties are high elasticity, stiffness, toughness, strength, good thermal resistance, and high chemical stability. This volume, Advances in Diverse Applications of Polymer Composites: Synthesis, Application, and Characterization, provides important information that will be valuable for researchers, postgraduate students, professors, and instructors working in the field of polymers"-- Provided by publisher.

Identifiers: LCCN 2022050442 (print) | LCCN 2022050443 (ebook) | ISBN 9781774910962 (hbk) | ISBN 9781774910979 (pbk) | ISBN 9781003300526 (ebk)

Subjects: LCSH: Polymeric composites.

Classification: LCC TA418.9.C6 A2924 2023 (print) | LCC TA418.9.C6 (ebook) | DDC 620.1/92--dc23/eng/20221220

LC record available at https://lccn.loc.gov/2022050442

LC ebook record available at https://lccn.loc.gov/2022050443

ISBN: 978-1-77491-096-2 (hbk)
ISBN: 978-1-77491-097-9 (pbk)
ISBN: 978-1-00330-052-6 (ebk)

About the Editors

Suji Mary Zachariah

Research Scholar, International and Inter University Centre for Nanoscience and Nanotechnology, Kottayam, India; Centre de Recherche Christian Huygens, Lorient, France

Suji Mary Zachariah is currently working as Research Scholar under the guidance of Prof. Sabu Thomas at the International and Inter University Centre for Nanoscience and Nanotechnology, Mahatma Gandhi University, Kottayam, India. She is also affiliated with the Centre de Recherche Christian Huygens, Lorient, France. She has expertise in polymer science. Her research interests are in the field of polymer-based nanocomposites for electromagnetic interference shielding application. Miss Zachariah mainly works on the development of absorption dominated materials from sustainable resources and fabrication of several biopolymer composites with improved mechanical, electrical and shielding properties.

Yang Weimin, PhD

Professor of Mechanical and Electrical Engineering, Beijing University of Chemical Technology, China; Director, Polymer Processing and Advanced Manufacturing Center

Yang Weimin, PhD, is a Professor of Mechanical and Electrical Engineering at Beijing University of Chemical Technology and directs the Polymer Processing and Advanced Manufacturing Center. He is also a Distinguished Professor of the Chang Jiang Scholars Program, China Ministry of Education. His research interest is polymer processing and advanced manufacturing technology, mainly focused on the green process of polymer production, plastics precision molding, energy saving in tire manufacturing, nanofiber electrospinning, enhanced heat transfer in polymer processing, etc. His research group has undertaken over 30 projects supported by the National Science and Technology Support Plan, the National Science Foundation of China, and an industrial company. Based on the research results, he has applied for more than 200 invention patents (82 items authorized to present), published eight books (including, in English, *Advances in Polymer Processing,* published by Woodhead in UK and CRC in the USA), and more than 300 journal papers. He has received numerous honors and recognitions, including two awards from the China National Science and Technology Progress, 10 China Provincial Awards, and the Hou Te-Pang Chemical Science and Technology Award. Dr. Yang holds a BS degree in Mechanical Engineering, and an MS and a PhD degree in Chemical Process Equipment from Beijing University of Chemical Technology. He was a postdoctoral fellow at Polymer Processing Yokoi Lab. of the University of Tokyo, Japan. In recent years, Dr. Yang has been serving on the editorial board of some important journals related to polymer processing and has also been selected as Vice Chairman of the Experts Committee of China Plastics Processing Industry Association and Advisor Expert of China Rubber Processing Industry Association.

Maciej Jaroszewski, PhD

Adjunct Professor, Wroclaw University of Science and Technology, Wroclaw, Poland

Maciej Jaroszewski, PhD, is currently working as an Adjunct Professor at Wroclaw University of Science and Technology, Wroclaw, Poland. He is the author and co-author of three chapters and more than 60 articles dealing with HV insulation, ZnO arresters, dielectric spectroscopy, and EMI shielding. He is the editor of four books. He is with the Faculty of Electrical Engineering at Wroclaw University of Science and Technology, Poland, as an assistant professor and is also the head of the High Voltage Laboratory. He is a member of the Polish Committee of Standardization. He is a reviewer in, among others, IEEE TDEI, IEEE Trans. on Power Deliv., and the Royal Society of Chemistry Journals. His current research interests include high voltage insulation and dielectric spectroscopy.

Sabu Thomas, PhD

Vice-Chancellor, Mahatma Gandhi University; Founder Director and Professor, International and Inter University Centre for Nanoscience and Nanotechnology, Kottayam, India; School of Energy Materials, Mahatma Gandhi University, Kottayam, Kerala, India

Sabu Thomas, PhD, is currently the Vice-Chancellor of Mahatma Gandhi University and the Founder, Director, and Professor of the International and Inter-University Center for Nanoscience and Nanotechnology. He is also a full professor of Polymer Science and Engineering at the School of Chemical Sciences, Mahatma Gandhi University, Kottayam, Kerala, India. Professor Thomas is an outstanding leader with sustained international acclaim for his work in nanoscience, polymer science and engineering, polymer nanocomposites, elastomers, polymer blends, interpenetrating polymer networks, polymer membranes, green composites and nano-composites, nanomedicine, and green nanotechnology. Professor Thomas has been conferred an Honoris Causa DSc from the University of South Brittany, France, and the University of Lorraine, France. Very recently, he was awarded the Foreign Fellow of the European Academy of Sciences (EurASc) and was listed on Stanford University's list of top 2% scientists in the world. He received many national and international awards, and he was published over 1,000 peer-reviewed research papers, reviews, and chapters. He has co-edited 80 books, and he is the inventor of more than five patents (granted – 4, filed – 11).

Contents

Contributors ... *xi*

Abbreviations ... *xiii*

Symbols ... *xxi*

Introduction ... *xxiii*

Preface ... *xxv*

1. **Applications of Polymeric Green Composites in the Biomedical Field: A Review ... 1**
   A. S. Dutta

2. **Characterization of Natural Particulates Filled and E-Glass Fiber-Reinforced Sandwich Polymer Composites ... 29**
   C. Balaji Ayyanar and K. Marimuthu

3. **A Review on Past and Future Aspects of Silica in Drug Delivery and Sensing Applications ... 47**
   Harpreet Kaur and Gagandeep Kaur

4. **Tamarind Kernel Powder, Its Derivatives, and Their Modification Through Grafting: An Overview ... 69**
   J. H. Trivedi, Ageetha Vanaamudan, and H. C. Trivedi

5. **Electrical Properties of $NA_2 PB_2 LA_2 W_2 TI_4 V_4O_{30}$ Ferroelectric Ceramic ... 95**
   Piyush R. Das, S. Behera, S. K. Mohanty, and Khusboo Agrawal

6. **Investigation of Dielectric and Ferroelectric Properties of PVDF/0.5Ba $(Zr_{0.2}Ti_{0.8})O_3$–$0.5(Ba_{0.7}Ca_{0.3})TiO_3$ Composite ... 105**
   Sakti Prasanna Muduli, S. Parida, S. K. Rout, and S. K. Mahapatra

7. **Structural and Frequency-Dependent Electrical Properties of Lead-Free $Na_2 Ba_2 La_2 W_2 Ti_4 Nb_4 O_{30}$ Ceramics ... 121**
   S. Devi, S. Behera, and T. Sahu

8. Advanced Materials for Electromagnetic Shielding 135
Suji Mary Zachariah, Ananthu Prasad, Avinash R. Pai,
Yves Grohens, and Sabu Thomas

9. Refabrication of Natural Polymeric Material Through Grafting: A Review 175
Sweta Sinha

10. Biodegradable PHAs: Promising "Green" Bioplastics and Possible Ways to Increase Their Availability 191
Tatiana Gr. Volova, Evgeniy G. Kiselev, Aleksey V. Demidenko,
Ekaterina I. Shishatskaya, and Sabu Thomas

Index.. 219

Contributors

Khusboo Agrawal
Department of Physics, Veer Surendra Sai University of Technology, Burla–768018, Sambalpur, Odisha, India

C. Balaji Ayyanar
Department of Mechanical Engineering, Coimbatore Institute of Technology, Coimbatore–641014, Tamil Nadu, India, E-mail: balajiayyanar2007@gmail.com

S. Behera
Department of Physics, School of Applied Sciences, Centurion University of Technology and Management, Bhubaneswar–752050, Odisha, India, E-mails: saubhagyalaxmi.behera@cutm.ac.in

Piyush R. Das
Department of Physics, Veer Surendra Sai University of Technology, Burla–768018, Sambalpur, Odisha, India, E-mail: prdas63@gmail.com

Aleksey V. Demidenko
Siberian Federal University, 79 Svobodnyi Av., Krasnoyarsk–660041, Russia; Institute of Biophysics SB RAS, Federal Research Center "Krasnoyarsk Science Center SB RAS," 50/50 Akademgorodok, Krasnoyarsk–660036, Russia

S. Devi
Department of Physics, Centurion University of Technology and Management, Odisha, India

A. S. Dutta
Maharashtra Institute of Technology, Aurangabad, Maharashtra, India, E-mail: aastha2006@gmail.com

Yves Grohens
Christian Huygens Research Center, Rue de Saint-Maudé, Lorient–56100, France

Gagandeep Kaur
Department of Chemistry, Punjabi University, Patiala–147002, Punjab, India

Harpreet Kaur
Department of Chemistry, Punjabi University, Patiala–147002, Punjab, India, E-mail: preetjudge@yahoo.co.in

Evgeniy G. Kiselev
Siberian Federal University, 79 Svobodnyi Av., Krasnoyarsk–660041, Russia; Institute of Biophysics SB RAS, Federal Research Center "Krasnoyarsk Science Center SB RAS," 50/50 Akademgorodok, Krasnoyarsk–660036, Russia

S. K. Mahapatra
Center for Physical Sciences, Central University of Punjab, Bathinda–151001, Punjab, India

K. Marimuthu
Department of Mechanical Engineering, Coimbatore Institute of Technology, Coimbatore–641014, Tamil Nadu, India

S. K. Mohanty
Department of Physics, M. H. D. College, Chhatia, Jajpur, Odisha, India

Sakti Prasanna Muduli
Department of Physics, C.V. Raman Global University, Bhubaneswar–752054, Odisha, India

Avinash R. Pai
School of Chemical Sciences, Mahatma Gandhi University, Kottayam–686560, Kerala, India

S. Parida
Department of Physics, C.V. Raman Global University, Bhubaneswar–752054, Odisha, India,
E-mail: sabyasachiparida023@gmail.com

Ananthu Prasad
International and Inter-University Center for Nanoscience and Nanotechnology (IIUCNN),
Mahatma Gandhi University, Kottayam–686560, Kerala, India

S. K. Rout
Department of Physics, Birla Institute of Technology, Mesra, Ranchi – 835215, Jharkhand, India

T. Sahu
School of Physics, Sambalpur University, JyotiViahr, Burla–768019, Odisha, India

Ekaterina I. Shishatskaya
Siberian Federal University, 79 Svobodnyi Av., Krasnoyarsk–660041, Russia; Institute of Biophysics SB RAS, Federal Research Center "Krasnoyarsk Science Center SB RAS," 50/50 Akademgorodok, Krasnoyarsk–660036, Russia

Sweta Sinha
Assistant Professor, Department of Applied Sciences, Amity University Jharkhand, Ranchi, Jharkhand, India, E-mail: sweta.sinha2203@gmail.com

Sabu Thomas
Siberian Federal University, 79 Svobodnyi Av., Krasnoyarsk–660041, Russia;
International and Inter-University Center for Nanoscience and Nanotechnology (IIUCNN),
Mahatma Gandhi University, Kottayam–686560, Kerala, India

H. C. Trivedi
Department of Chemistry, Faculty of Applied Sciences, Parul University, Limda, Vadodara–391760, Gujarat, India

J. H. Trivedi
Post Graduate, Department of Chemistry, Sardar Patel University, Vallabh Vidyanagar–388120, Gujarat, India, E-mail: drjignesh2575@yahoo.co.in

Ageetha Vanaamudan
Department of Chemistry, Faculty of Applied Sciences, Parul University, Limda, Vadodara–391760, Gujarat, India

Tatiana Gr. Volova
Siberian Federal University, 79 Svobodnyi Av., Krasnoyarsk–660041, Russia; Institute of Biophysics SB RAS, Federal Research Center "Krasnoyarsk Science Center SB RAS," 50/50 Akademgorodok, Krasnoyarsk–660036, Russia, E-mail: volova45@mail.ru

Suji Mary Zachariah
International and Inter-University Center for Nanoscience and Nanotechnology (IIUCNN), Mahatma Gandhi University, Kottayam–686560, Kerala, India, E-mail: sujimaryzachariah@gmail.com

Abbreviations

^{13}C NMR	carbon-13 (C-13) nuclear magnetic resonance
^{1}H-NMR	proton nuclear magnetic resonance
3D	three-dimensional
AAm	acrylamide
ACL	anterior cruciate ligaments
Ag	silver
AIBN	azobisisobutyronitrile
AIE	aggregation-induced emission
$AlCl_3$	aluminum chloride
AN	acrylonitrile
APS	ammonium persulphate
APTES	3-aminopropyl-triethoxysilane
ATP	adenosine triphosphate
ATRP	atom transfer radical polymerization
AuNPs	gold nanoparticles
$BaTiO_2$	barium titanate
BC	bacterial cellulose
BET	Brunauer-Emmett-Teller
BPO	benzoyl peroxide
bpy	2,2′-bipyridine
BZT-BCT	0.5Ba(Zr0.2Ti0.8)O3–0.5(Ba0.7Ca0.3)TiO3
$CaCl_2$	calcium chloride
CAN	ceric ammonium nitrate
CCB	conductive carbon black
CDM	cell samples dried
CF	carbon fibers
CFR	carbon-fiber-reinforced
ChNC	chitin nanocrystals
CIC	critical incorporation concentration
CMCs	ceramic matrix composites
CMT	carboxymethyl tamarind
CMT-g-PAM	carboxymethyl tamarind-graft-polyacrylamide
CMTKP	carboxymethyl tamarind kernel polysaccharide

CMTKP-g-PAN	carboxymethyl tamarind kernel powder-graft-polyacrylonitrile
CNC	cellulose nanocrystals
CNF	carbon nanofibers
CNP	cellulose nanopaper
CNS	carbon nanotube structures
CNT	carbon nanotubes
CNW	carbon nanowires
Co-Cr and Ti alloys	cobalt-chromium and titanium alloys
CR	Congo red
CTAB	cetyltrimethylammonium bromide
CuBr/bpy	copper bromide/Bipyrridine
dB	decibels
DBSA	dodecylbenzenesulfonic
DLS	dynamic light scattering
DMF	N,N-dimethyl formamide
DS	degree of substitution
DSC	differential scanning calorimetry
DTA	differential thermogravimetry analysis
EA	ethyl acrylate
EBT	erichrome black T
ECM	extracellular matrix
EDS	energy dispersive x-ray spectroscopy
ELP	electroless plating
EM	electromagnetic
EMA	ethyl methacrylate
EMI	electromagnetic interference
EMT	effective medium theory
FESEM	field emission scanning electron microscope
FFA	free fatty acids
FITC	fluorescein isothiocyanate
FTIR	Fourier transforms infrared
GCE	glassy carbon electrode
GC-MS	gas chromatography-mass spectrometry
GF	graphene foam
GGM	galactoglucomannan
GM	guar galactomannan
GNR	graphene nanoribbons

GO	graphene oxide
GPC	gel permeation chromatography
GSM	gram per square meter
GTS	grafted tamarind seed polysaccharide
GTSP	grafted TSP
HDPE	high-density polyethylene
HEMA	hydroxy ethyl methacrylate
HF	high frequency
hMSC	human mesenchymal stem cells
H-Na-PCMTKP-g-PAN	partially hydrolyzed sodium salt of partially carboxymethylated tamarind kernel powder-graft-polyacrylonitrile
HPSEC-MALLS	high-performance size-exclusion chromatography-Multiangle laser light scattering
HR-SEM	high-resolution scanning electron microscope
ICPs	intrinsically conduct polymers
ISF	*in-situ* nanofiber
KOH	potassium hydroxide
LC-MS/MS	liquid chromatography-mass spectrometer
LDPE	low-density-polyethylene
LFs	leather fibers
LOD	limit of detection
LSNs	luminescent silica nanoparticles
LSW	leather solid waste
MA	methyl acrylate
MAAc	methacrylic acid
MB	methylene blue
MCPE	modified carbon paste electrode
MIP	molecularly imprinted polymer
MMA	methyl methacrylate
MMCs	metal matrix composites
MMP	matrix metalloproteinase
MMT	montmorillonite
M_n	weight average molecular weight
MPa	megapascal
MPB	morphotropic phase boundary
MRI	magnetic resonance imaging
MSNs	mesoporous silica nanoparticles

MSW	municipal solid wastes
MW	microwave
M_w	number average molecular weight
NaAlg	sodium alginate
NaCl	sodium chloride
Na-PCMTKP	sodium salt of partially carboxymethylated tamarind kernel powder
Na-PCMTKP-g-PAN	sodium salt of partially carboxymethylated tamarind kernel powder-graft-polyacrylonitrile
Na-PCMTKP-g-PMMA	sodium salt of partially carboxymethylated tamarind kernel powder-graft-polymethyl methacrylate
Na-PCMTKP-PEMA	sodium salt of partially carboxymethylated tamarind kernel powder-graft-polymethyl methacrylate
NMR	nuclear magnetic resonance
NPLWTV	$Na_2Pb_2La_2W_2Ti_4V_4O_{30}$
NTCR	negative temperature co-efficient of resistivity
NVP	N-vinyl pyrrolidone
ODTMS	octadecyltrimethoxysilane
P	phosphor
P(3HB)	poly-3-hydroxybutyrate
pAb	polyclonal antibody
PAH	poly allylamine hydrochloride
PAM	polyacrylamide
PAM-g-TSG	polyacrylamide grafted tamarind seed gum
PAM-g-TSP	polyacrylamide-grafted-tamarind seed polysaccharide
PAN	polyacrylonitrile
PANI	polyaniline
PANI/SrM/MWCNT	polyaniline/strontium ferrite/multiwalled carbon nanotubes
PC	plain cement
PCB	printed circuit boards
PCL	polycaprolactone
PDLLA	poly D,L-lactic acid
PE	polyethylene
PEDOT	poly(3,4-ethylene dioxythiophene)

PEEK	polyether ether ketone
PEG	polyethylene glycol
PEM	polyelectrolyte multilayers
PET	polyethylene terephthalate
PGA	polyglycolic acid
PHA	polyhydroxyalkanoates
PHEMA	poly (2-hydroxyethyl methacrylate)
PLA	polylactic acid
PLGA	polylactic co-glycolic acid
PLLA	poly L-lactic acid
PMCs	polymer matrix composites
PMMA	polymethyl methacrylate
POFA	palm oil fuel ash
PP	polypropylene
PPy	polypyrrole
PSS	polystyrene sulfonate
PT	polythiophene
PTFE	polytetrafluorethylene
PU	polyurethane
PVA	polyvinyl alcohol
PVAc	polyvinyl acetate
PVDF	glass transition relaxation
PZT	$Pb(Zr,Ti)O_3$
RCIM	Russian collection of industrial microorganisms
RF	radio frequency
rGO	reduced graphene oxide
RH	rice husk
RHA	rice husk ash
RM	red mud
RS	rice straw
RSA	rice straw ash
SE	shielding effectiveness
SE_A	shielding effectiveness by absorption
SEM	scanning electron microscope
SE_M	shielding effectiveness by multiple internal reflections
SE_R	shielding effectiveness by reflection
SE_T	shielding effectiveness

SI-ATRP	surface-initiated atom transfer radical polymerization
SiO_2	silicon dioxide
SLS	static light scattering
SNA	scalar network analyzer
SPR	smoke production ratio
ssDNA	single-stranded DNA
SSP	single-source precursor
SU	simulated urine
TAM	tamarind mucilage
TAM-g-PAM	tamarind mucilage-graft-polyacrylamide
TB	tungsten bronze
T_d	thermal degradation temperature
TEM	transmission electron microscopy
TEOS	tetraethyl orthosilicate
TF	transfemoral
TG	tamarind gum
TGA	thermogravimetry analysis
TiO_2	titanium dioxide
TKP	tamarind kernel polysaccharide
TKP	tamarind kernel powder
TKP-g-PAM	tamarind kernel powder-graft-polyacrylamide
T_m	melting temperature
TSG	tamarind kernel gum
TSP	tamarind seed polysaccharide
TSP	tamarind seed powder
TSP	total smoke production
TT	transtibial
TTA	thenoyltrifluoroacetone
UHMWPE	ultra-high molecular weight polyethylene
UHPC	ultra-high performance concrete
UV	ultra-violet waves
V	volume
VNA	vector network analyzer
WPC	wood/polyethylene composites
WPC_{CTRL}	wood polyethylene composites without CCB coating
$WPC_{H\text{-}CCB}$	honeycomb-distributed WPC

WR	woven roving
XG	xyloglucan
XRD	X-ray diffraction
ZnO	zinc oxide

Symbols

μ	specific growth rate
°C	degree Celsius
μm	micrometer
A	surface area of the sample
A°	Armstrong
C_0	geometrical capacitance
C_x	degree of crystallinity
E	applied electric field
Ec	coercive field
eV	electron volt
g/cm^3	gram per cubic centimeter
K	degradation rate constant
K	Kelvin
KHz	kilohertz
K_i	inhibition constant
K_s	saturation constant
M(ω)	complex electrical modulus
M_0	starting mass of the specimen
MHz	megahertz
M_i	mass of the specimen after degradation
N/mm^2	Newton per millimeter square
P	polarisation
P	volumetric productivity
P_{max}	maximum polarisation
R_b	bulk resistance
RPM	rotations per minute
S	substrate concentration
tanδ	dielectric loss
V_f	volume fraction (percentage)
W_f	weight fraction
X_c	catalytically active biomass
X_c	crystallinity

X_{total}	total biomass
Y	yield coefficient of the polymer
Y^*	complex admittance
$Z(\omega)$	complex impedance
γ	diffusivity
ε^*	complex permittivity
ε_i	dielectric constant of filler
ε_m	dielectric constant of matrix
θ	Braggs angle
λ	wavelength
σ	conductivity
ω	angular frequency

Introduction

The word "polymer" is a Greek word meaning "many parts," which is made from the same or different monomers. Lots of synthetic and natural polymers for various applications are available on the market, and most of the chemical industries are devoted to the production of synthetic polymers. Polymers, whether natural or synthetic, have a wide variety of applications ranging from household applications to biomedical and other high-end applications. This book covers the synthesis, characterization, and property study of some of these polymers, including their morphology, structure, and dynamics. It also introduces the most recent innovations and applications of polymers, fillers, and composites in the electronics, biomedical, pharmaceutical, and engineering industries.

Chapters 1, 2, 3, 4, 9, and 10 mainly deal with natural polymers and fillers, while Chapters 5 and 7 deal with ceramics. Chapter 8 extensively discusses numerous polymers for radiation shielding applications.

Preface

Advances in Diverse Applications of Polymer Composites: Synthesis, Application, and Characterization is a book that focuses on polymers, their characterization, and their various applications. These include drug delivery applications, electromagnetic (EM) shielding, ferroelectric applications, and many more.

This book is the materialization of the effort of a myriad of authors, editors, and reviewers, and we wish you all enjoy this book as much as we are thrilled while producing it.

Wishing you a pleasant reading.

—Editors

CHAPTER 1

Applications of Polymeric Green Composites in the Biomedical Field: A Review

A. S. DUTTA

Maharashtra Institute of Technology, Aurangabad, Maharashtra, India, E-mail: aastha2006@gmail.com

ABSTRACT

Composites have been used by human beings for quite a long time. But a study of its properties, variations in its grade, and its field of applications have gained importance in recent years. But the major drawback of these traditional composites is that they are mostly based on either synthetic materials like polymers, nano-materials, or mining resources like metals. These materials pose hazards to the environment in some or the other way. So, in recent days the focus has been on materials from the environment which can be easily returned back to the environment. Green composites are defined as the bio-composites, in which a bio-based polymer matrix is reinforced with the help of natural fibers such as cotton, wool, etc., and they represent an emerging area in the field of polymer science. The sources of these composites are biodegradable, renewable materials. These composites are gaining importance in various fields, including the biomedical field. Some applications of these composites in the biomedical field will be discussed in this chapter.

Advances in Diverse Applications of Polymer Composites: Synthesis, Application, and Characterization.
Suji Mary Zachariah, Yang Weimin, Maciej Jaroszewski, & Sabu Thomas (Eds.)

1.1 INTRODUCTION

Polymeric composites have gained importance in various fields viz: automobile, construction, electronics, packaging, and biomedical field [1]. Mostly in the clinical field, many naturally occurring fibers find their applications. Other synthetic materials that are being traditionally used are metals, glass, ceramics, polymers, etc. The most widely used polymer is polylactic acid (PLA) which is obtained from casein and corn, which are naturally found materials. The properties expected from these materials are biocompatibility, controlled rate of degradation, and nontoxicity. Other natural polymers that normally find their usage in the biomedical field are polysaccharides or proteins, polycaprolactone (PCL), and polyglycolic acid (PGA). Some water-soluble polymers being used are polyvinyl alcohol (PVA), polyvinyl acetate (PVAc), guar gum, etc. [2].

1.2 IMPORTANCE OF GREEN COMPOSITES

When the resin and fibers of green composites are discarded in the environment, they are acted upon by microorganisms, and they get biodegraded and get converted to carbon dioxide and water, which get absorbed by the plants [3]. Thus, they help us to keep the environment safe and clean. So, in today's polluted world, their demand is growing very rapidly.

1.3 CLASSIFICATION OF COMPOSITES [4]

Composite materials are classified based on two parameters:

- Composite materials depending on the material used as a matrix; and
- Composite materials depend upon the structure of the material used for reinforcement.

On the basis of the matrix phase, composites are normally categorized as:

- Metal matrix composites (MMCs);
- Ceramic matrix composites (CMCs);
- Polymer matrix composites (PMCs).

According to the various types of reinforcement, composites can be classified as:

- Particulate composites;
- Fibrous composites;
- Laminate composites.

The fibrous composites are further subdivided depending on the source of fiber as:

- Natural fiber (bio-fibers);
- Synthetic fiber.

The bio-fibers-based composites based on the nature of the matrix are classified as:

- Non-biodegradable matrix;
- Biodegradable matrix;
- Green composites are those bio-composites that comprise natural/ bio-fiber and biodegradable polymers.

1.3.1 NATURAL FIBER COMPOSITES

As the name suggests, these fibers are obtained from natural sources and are reinforced into the polymeric matrix to form natural fiber composites. Their various sources are animals, minerals, plants, etc. [4, 5].

1.3.2 ANIMAL FIBERS

Animal fibers generally are rich in protein content. The well-known examples are, for example, mohair, wool, and silk. Avian fibers are the fibers that are collected from birds feathers. Various sources of animal fibers are shown in Figure 1.1.

1.3.3 MINERAL FIBERS

Mineral fibers are also natural fibers procured from minerals. Examples are asbestos fibers, ceramic fibers, and metal fibers.

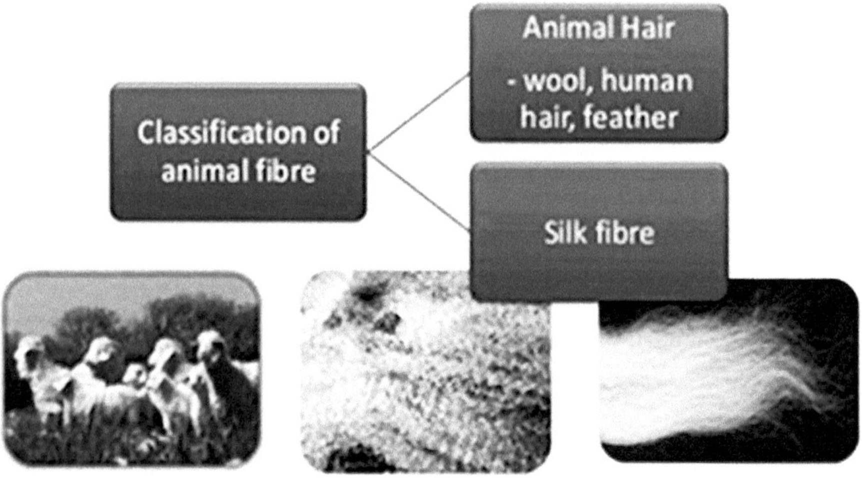

FIGURE 1.1 Various sources of animal fibers.

1.3.4 PLANT FIBERS

Plant fibers are rich in cellulose content, e.g., cotton, jute, hemp, ramie, sisal, flax, etc. These fibers have varieties of extraction sources, such as: (i) fibers extracted from fruits, for example, coconut fiber; (ii) fibers found in stems of plants, for example, bast fiber; and (iii) fibers found in leaves of plants/trees, for example, sisal.

Various sources of plant fibers are shown in Figure 1.2.

1.4 CLASSIFICATION OF GREEN COMPOSITES

Green composites are broadly classified on type of reinforcement and polymer materials as:

1. **Totally Renewable Composites:** Here, the matrix, as well as the reinforcing material, are obtained from renewable resources.
2. **Partly Renewable Composites:** Here, the matrix is extracted from renewable resources, whereas reinforcement fiber is a tailor-made material.
3. **Partly Renewable Composites:** Here matrix used is of synthetic origin, but reinforcement fiber is a natural biopolymer [1, 7].

FIGURE 1.2 Various sources of plant fibers.

Source: Reprinted from Ref. [55].© 2022 by authors and Scientific Research Publishing Inc.

1.5 FACTORS AFFECTING PROPERTIES OF GREEN COMPOSITES

1.5.1 INTERFACIAL ADHESION

In composite formation, the ingredients involved are the polymer matrix, reinforcing element (fiber), and the fiber and matrix interface. For attaining superior properties, the interfacial adhesion bonding of fiber and matrix should be very strong [4]. This interfacial adhesion is attained by means of some chemical reaction or by the adsorption of matrix molecules on the surface of the fiber.

1.5.2 SHAPE AND ORIENTATION OF REINFORCING PHASE

In composites, the reinforcing materials or the dispersed phase are available in various shapes like particles, flakes, fibers, and laminates [4]. The main aim of adding the reinforcing material is to improve properties, especially mechanical properties and also make the composites available at an economical rate. The reinforcing material plays a very crucial role in improving the thermal expansion coefficient and the conductivity of the composites in which they are incorporated [6]. As the particle size and shape change, the properties of composites change drastically. Some of the regularly available shapes of reinforcing materials are spherical, cubic, platelet, and regular or irregular geometry [4]. The various shapes of reinforcing fillers are depicted in Figure 1.3.

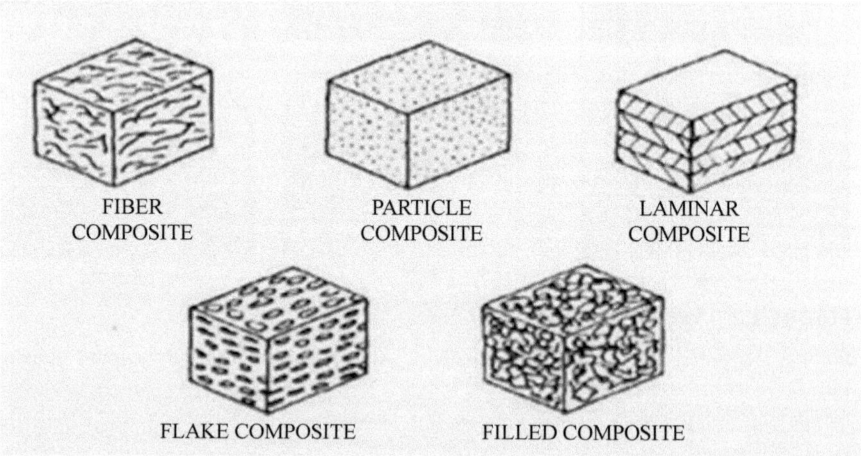

FIGURE 1.3 Various shapes of reinforcing fillers.

1.5.3 ROLE OF MATRICES

Matrices are mostly viscous and resinous in nature. Compared to reinforcing materials, their properties and inferior, but they play a very important role in composite formation for the following reasons:

- Imparting proper shape to the composite;
- Keeping the fibers aligned in proper places;
- Transferring the stresses that are applied to the fibers;
- Protecting the reinforcement from various environmental factors, such as chemicals and moisture;
- Protecting the surface of fibers from mechanical degradation.

1.5.4 PERFORMANCE OF COMPOSITES

The overall property of a composite is reflected by the individual characteristic properties of reinforcing fillers and the matrix. The factors dictating the properties of fibers are discussed in subsections.

1.5.4.1 LENGTH OF THE FIBER

The fibers incorporated are of various lengths, viz: long, short, or continuous. For ease of processing and proper orientation, long, continuous fibers are preferred as compared to short fibers. Other benefits of using long fibers against short length fibers include properties like high strength, impact resistance, low shrinkage percentage, enhanced surface finish, and dimensional stability. Short-length fibers, on the other hand, also provide easy workability, faster fabrication cycle times, and are also more economical than long fibers [6]. To achieve the isotropic behavior of the composites, short fibers which are randomly oriented can be added.

1.5.4.2 ORIENTATION

Normally unidirectional orientation of the fibers provides the composites with high stiffness and strength in the orientation direction. On the other hand, if the fibers are aligned in multi directions, such as in a mat, there will be property enhancement in the direction in which the fibers are aligned [1, 6].

1.5.4.3 SHAPE

Mostly for ease of handling and manufacturing, fibers are mostly circular in shape. But for specific applications, fibers may be added in other shapes viz: square and rectangular also.

1.5.4.4 MATERIAL

The material of the fiber has a direct effect on the properties of the composite, especially mechanical properties. Normally it is expected that the fibers should have high elastic moduli and strength as compared to the matrix [1, 6].

1.6 DESIRED PROPERTIES FOR BIOMEDICAL APPLICATIONS

Any material which finds application in the biomedical field should satisfy certain criteria as it is in direct contact with the human body. The factors to be considered are:

- Material should be non-toxic;
- It should be bio-degradable;
- It should be bio-compatible.

In addition to these factors, bio-composite material should also exhibit optimum superhydrophobicity, adhesion, and self-healing property.

1.6.1 SUPER HYDROPHOBICITY

Materials that exhibit superhydrophobic surfaces have very tough wettability properties. This is seen in many insects and plants. This property is very useful in biomedical applications because it reduces the chances of blood coagulation due to unfavorable platelet adhesion [8, 9].

1.6.2 ADHESION

This is another phenomenon that is widely seen in plants and animals because it is a basic criterion for the survival of organisms. Because of this property, organisms can attach themselves to their host either permanently or temporarily. Some tailor-made polymers exhibiting this property have been developed and are being used in biomedical applications [9].

1.6.3 SELF-HEALING

Our human body is made up in such a manner that whenever there is an injury, it automatically repairs or replaces the damaged tissues; this concept is known as self-healing. But self-healing is possible only if the portion of the injury is not too great. If the level of injury or damage is beyond the capacity of self-healing, then the requirement for an alternative material arises, which in medical terms is known as an implant. Implants, wherever replaced in the human body, are subjected to various forms of load, wear, aging, etc., and may result in failure, and again, the requirement of replacement arises. For these reasons, researchers are trying to develop self-healing materials. Normally they are categorized as first and second-generation materials. The first generation of self-healing bio-materials is of irreversible nature in which the composites get irreversibly repaired without restoring the damaged matrix, while in the second generation reversible mechanism occurs, and the damaged matrix is restored. Some typical properties (Figure 1.4) expected from self-healing polymers are:

- It should have the ability to heal the damaged parts of materials several times;
- It should have the ability to heal the materials automatically;
- Reduced maintenance cost;
- It should have the ability to heal defects of any size;
- It must show equal or better performance as compared to the traditional materials;
- It should be more economical than the materials being used regularly [8, 10].

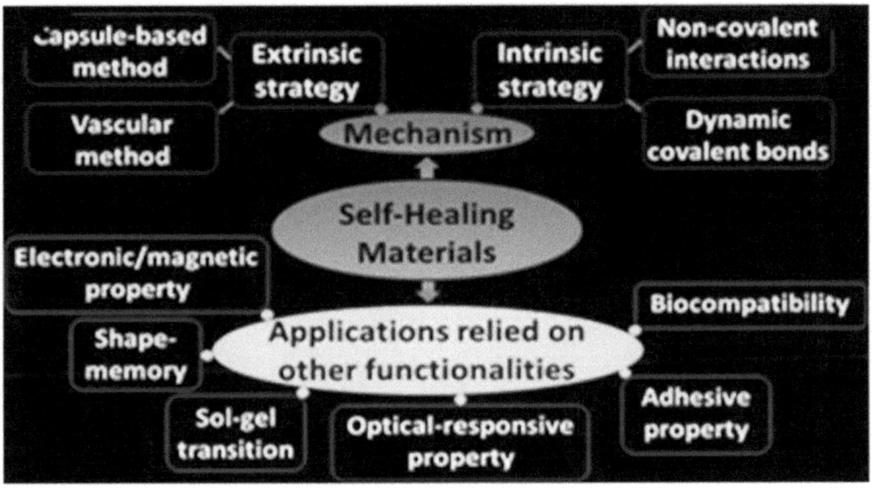

FIGURE 1.4 Characteristics of self-healing materials.

1.7 APPLICATIONS OF POLYMERIC GREEN COMPOSITES IN THE BIOMEDICAL FIELD

A composite is a commonly used material in the medical field. Due to recent advancements in medical science as well as composites manufacturing, these composites are available for usage in various ways. The medical field today utilizes a vast number of devices and implants, whereas composites are being used frequently. Composites are being used in various forms such as sutures, bone, and joint replacements, vascular grafts, heart valves, intraocular lenses, dental implants, pacemakers, biosensors, artificial hearts, etc., for replacing and restoring the function of disturbed or degenerated tissues or organs as well as for improving the organ functioning [12, 44, 45]. Fields of application of biomedical polymers are schematically shown in Figure 1.5.

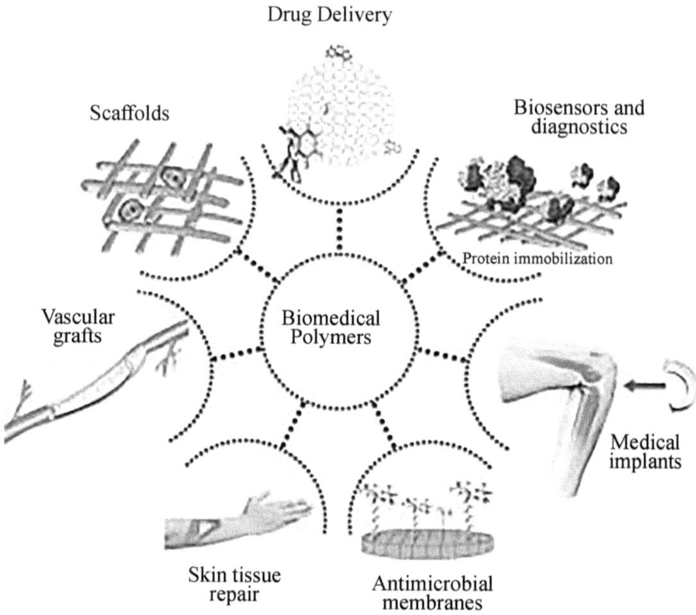

FIGURE 1.5 Fields of application of biomedical polymers.

Source: Reprinted from Ref. [56]. © 2022 Technology Times.

1.7.1 TISSUE ENGINEERING

Tissue engineering basically involves the regeneration of tissues and organs and the improvement of their biological functions. Its basic concept is to combine a biodegradable matrix (scaffold), living cells, and/or biologically active molecules to form a structure that will help in promoting the repair and regeneration of the targeted tissue. Composites are finding major applications in this field, and therefore, many innovative materials are coming up to serve the purpose. As a result, biodegradable materials like poly l-lactic acid (PLLA) and poly d,l-lactic acid (PDLLA) have often been combined with other degradable polymers such as poly-lactic co-glycolic acid (PLGA), polyethylene glycol (PEG), and chitosan to form composites with desired material properties [11]. Chitosan, which is derived from N-deacetylated chitin, shows properties like bioresorption, absence of cytotoxicity, and low environmental impact during processing. Accordingly, chitosan and its composites have found numerous applications in the biomedical field, especially in the three-dimensional (3D) scaffolds application in which cells are seeded before implantation [12, 13], because of its various enhancing properties such as biocompatibility, biodegradability, and osteoconductivity. The tissue engineering paradigm is shown in Figure 1.6. For surfaces and scaffolds for cartilage cell growth, glycolic acid in fibrous form (PGA), as well as lactic-co-glycollic acid (PLGA) also in fibrous form, are being used. When the cell is implanted at the defective cartilage site, it results in bulk matrix production of the implanted polymeric material. The properties of the fibers that affect the quality of the neocartilage that is being constructed are: fiber diameter, interfiber distance, and biodegradation rate. For replacing damaged vascular grafts, highly permeable polyester woven-fiber tubes are being used, but this implantation results in severe blood leakage through the walls of the vascular graft. This leakage problem increases the pre-clotting time of blood, and therefore, the grafts are coated with various impermeable materials to form a composite. The matrix acts in two manners: as a sealant as well as a load-bearing material. For example, naturally degradable composites used in vascular prostheses are composed of alginate and gelatin. Here Dacron fibers are used for reinforcement, and the matrix is composed of alginate and gelatin. These composites help in the sealing of the vascular prosthesis [5].

In recent years nanofibrous scaffolds have been targeted as composite materials because they have the physical properties of artificial polymers and the bioactivity property of natural polymers. The scaffold, which is made out of composites, must possess certain characteristics such as high porosity, high surface area, structural strength, specific 3D shape, and the most important one, that is biodegradability [11]. 3D hydroxyapatite/chitosan nanocomposite rods having layered structures have been constructed via in situ precipitation method. This resulted in bioabsorbable parts that can be used for bone fracture internal fixation [16, 17]. Zirconium-chitin-based composites have been used for preparing 3D biocompatible scaffolds. This has been introduced by Ehrlich et al. [18].

Such scaffold specimens were prepared for the first time using the hydrothermal synthesis method. These zirconium-chitin composites are being used for a broad range of applications ranging from synthetic catalysts to biocompatible materials for bone and dental repair [18].

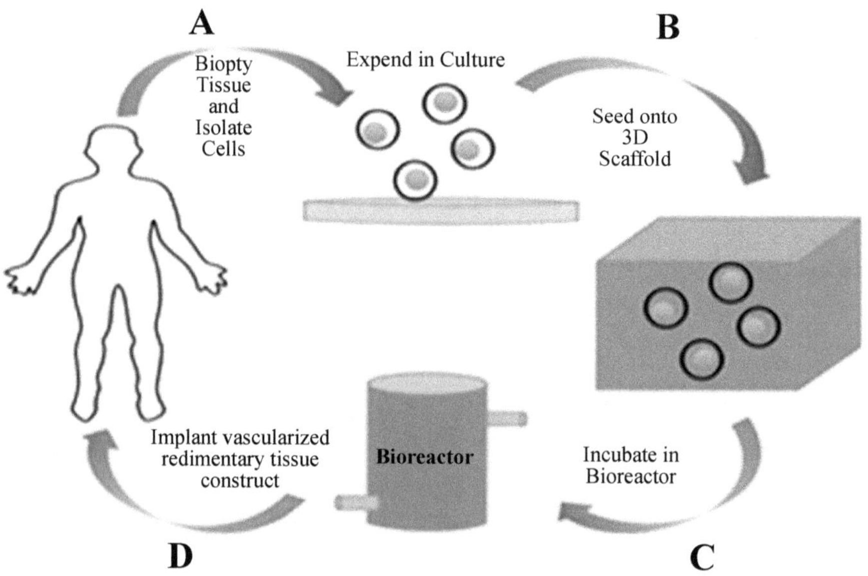

FIGURE 1.6 Tissue engineering paradigm.

Silk contains a high content of protein and is very easily susceptible to proteolytic degradation. Also, over a longer period of time, it gets slowly absorbed. Normally, the degradation rate of implants mainly depends on the physiological status of the patient, the mechanical environment of the implantation site, and the types and dimensions of the silk fibers. The slow rate of degradation of silk in vitro and in vivo makes it a very useful material to be used as biodegradable scaffolds which are having slow tissue ingrowths. This is necessary because the biodegradable scaffolds should not only get retained at the implantation site but also maintain their mechanical properties and support the growth of cells until the regenerated tissue is capable of fulfilling the desired functions. The degradation rate should be maintained such that it is at par with the rate of neo-tissue formation, so that it can compromise with the load-bearing capabilities of the tissue [19].

Meinel et al. mainly focused on cartilage tissue engineering by using silk protein scaffolds and reported that these silk scaffolds are mostly suitable for tissue engineering of cartilages which are starting from human mesenchymal stem cells (hMSC). These are derived from bone marrow, and have high porosity, slow degradation, and good structural integrity [20]. Recent research on silk has resulted in the development of a wire rope matrix, which is used for the development of autologous tissue-engineered anterior cruciate ligaments (ACL) using a patient's own adult stem cells [21].

Cross-linked hydrogel networks have high water retaining capacity and excellent barrier properties, and are therefore suitable for being used as a scaffold for skin regeneration. But these scaffolds lack their mechanical properties. To overcome this drawback, an attempt was taken to enhance these properties viz: tensile and breaking strength of the hydrogel, which was composed of poly (2-hydroxyethyl methacrylate) (PHEMA). In this hydrogel polyurethane (PU) fibers (Spandex) were used as the reinforcing material. This resulted in hydrogels with excellent mechanical properties [5]. Normally, high compressive strength and good shear properties are desirable in cartilage repair. For this purpose, a composite was designed using the molten linear low-density grade of polyethylene which was coated on top of woven 3D fabric made up of ultra-high molecular weight polyethylene. This composite showed mechanical properties nearly the same as that of the natural cartilage [22]. For hemodialysis application, the graft used should be resilient to frequent needle punctures. A method used for grafting involves a sandwich design in which Polytetrafluoroethylene fibers are placed in between the layers of porous expanded polytetrafluoroethylene. When the needle is pricked in the body, the fibers are pushed aside, and after it is withdrawn, the fibers create a baffle effect which results in a reduction of blood leakage and improves resealing [23].

1.7.2 IMPLANTATION

Another field where green composite is being widely preferred in the biomedical field is the field of implants and medical devices. It is expected that after a stipulated period of time, these will get dissolved and absorbed in the human body. Proper implant design and implantation technique assure that the desired physicomechanical properties of the fixations are achieved mainly by avoiding devascularization and additional tissue trauma at the site of the fracture. The most common materials that are used in these implants for medical applications are generally metals, for example, stainless steel, cobalt-chrome, and titanium alloys, polymers such as ultrahigh molecular weight polyethylene, polyether-ether-ketone, and ceramics such as hydroxyapatite. However, this trend has shifted in the last two decades of the 20th century. At present bioabsorbable or biodegradable bio-materials are preferred to biostable biomaterials for medical devices as they help the body not only to repair but also regenerate the damaged tissues [11]. Biodegradable materials should satisfy an important property of degrading either hydrolytically or enzymatically.

Composite materials have found wide usage in the orthopedic field, particularly in bone fixation plates, hip joint replacement, bone cement, and bone grafts. Fiber composites can be easily tailored to match the specific mechanical properties of the adjoining bone. Carbon-fiber composites having a matrix of Polyether ether ketone (PEEK) or polysulfone can be fabricated, and it shows excellent stiffness and tensile strength properties. But their fabrication is very difficult, and also, they have very less durability. But because of the inherent advantages of tailor ability, flexibility, and non-corrosiveness, their application is not yet restricted [5].

Problems arising because of the biocompatibility of the above composite materials are basically because of their carbon debris. This problem is normally avoided by painting the composites by using minerals such as hydroxyapatite or alloys of titanium. Examples of composite bone plates include partially adsorbable laminate, which is made up of continuous carbon fiber in a polylactide (PLA) matrix. Fully resorbable composites are made by adding calcium-phosphate glass fibers in PLA matrix. Continuous poly (L-lactide) fibers in a PLA matrix also produced a fully resorbable composite. But all the above three grades of composites, lack in their mechanical properties and degrade very rapidly. For usage in bone plates non-resorbable materials showing good impact and fatigue resistance are normally given preference. Such bone plates are mostly made up of a carbon-epoxy mixture [5].

1.7.2.1 BONE FRACTURE REPAIR

Skeletal bone is made up of collagen fibers which are composed of hydroxyapatite nanocrystals precipitate. Bone fractures can be treated basically in two different ways viz: internal fixation and external fixation. In internal fixation, the fracture site is opened up, whereas in external fixation, it is not required. While carrying out the external fixation, bone fragments are aligned in various manners such as splints, braces, casts, and external fixators [24]. For these alignments, the materials used are either plaster bandages or casting materials [25]. Composite materials made out of woven cotton fabrics (woven gauze) and plaster of Paris matrix (calcium sulfate), fabrics of glass, and polyester fibers are mostly chosen here. It is expected that an ideal cast material could be easily handled, conformable to the anatomical shapes, light in weight, strong, stiff, hydrophobic, air-permeable, and easily removable. External fixators are prepared by carbon fibers (CF)/epoxy composite as they are light in weight, have sufficient stiffness and strength. The bone fixation made of polymer composites are assessed by means of radiography, and they do not create artifacts in the radiographs. In the internal fixation method, the bone fragments are positioned together in different ways using various implants, such as intramedullary nails, wires, pins, screws, and plates [24]. The simplest form of implantation is surgical wire and pins, which are used for joining small bone fragments. They are also helpful in providing additional stability when long bones get fractured, for example, humerus, femur, tibia, radius, ulna, and fibula [26]. The non-resorbable composite plates are usually fabricated from either thermoplastic composite or thermosetting polymer composites. Epoxy-based composites have lost their importance due to the toxic effects of their monomers. Presently bone plates are being manufactured using CF with thermoplastics like polypropylene (PP), polyethylene (PE), polymethyl methacrylate (PMMA), Nylons, etc. The carbon fiber/PEEK composite shows the best properties like biocompatibility, fatigue resistance, no carcinogenicity, radiation resistance, and hydrolytic degradation. For osteosynthesis, carbon fiber/carbon and carbon fiber/polyetheretherketone (PEEK) composite screws were developed.

Non-resorbable polymer composites which were used for implantation were developed in such a way that they didn`t show any change in their stiffness during the entire implantation period. During the bone healing process, the bone is subjected to a gradual increase in stress to reduce the stress-shielding effect. This was proven by the loss in the rigidity of the plate as time progressed. At present, resorbable polymers are being more targeted for bone plate development. As the bio-resorbable polymers lose most of their mechanical properties in a couple of weeks, few researchers have proposed fully resorbable composites composed of PLA fibers and calcium phosphate-based glass fibers. The advantages of resorbable composites-based implants are that their removal is not needed. However, their applications were restricted only to implants, which were subjected to moderate load-bearing capacity. To improve the mechanical properties, carbon and polyamide fibers were added in the resorbable composites and then categorized as partially resorbable composites. The resorbable polymers were modified with a variety of non-resorbable materials, including carbon and polyamide [27, 28].

Intramedullary nails or rods are generally used for stapling long bone fractures, such as femoral neck bone. It is introduced into the intramedullary cavity of the bone, and its position is fixed either with the help of screws or with a friction fit method. The intramedullary nail fixation shows poor torsional resistance than plate fixation, but its bending resistance is better than plate fixation. Most preferred implant was Stainless steel. But recently, Lin et al. has suggested for a short glass fibers and PEEK-made composite material for intramedullary application. Few researchers have also developed a liquid crystalline polymer composite, which is reinforced with unidirectional carbon fiber for intramedullary rod. This composite is biologically inert, and has high flexural strength material and elastic modulus close to the bone [24, 29].

1.7.2.2 SPINAL IMPLANTATION

The spine of the human body acts as a strong, central axis onto which a load of an entire skeleton gets distributed. It protects the spinal cord as well as the roots of delicate nerves, which are connected to the peripheral nerves of the brain [24, 30]. Normally disorders related to the spine include the metastasis of the vertebral body as well as the disc, facet degeneration, disc herniation, and stenosis. Some structural abnormalities like scoliosis, kyphosis, and spondylolisthesis are also commonly seen. Spinal fusion and replacement of the disc are normally used for spine disorder treatment.

One method of spinal fusion removes the affected portion of vertebrae and replaces it with a synthetic bone graft.

Bioglass/polyurethane (PU) composite has been developed for vertebral body bone grafting [29, 31]. Another method involves the usage of some specific vertebral prostheses, for example, cages, baskets, and threaded inserts, which are composed of either metals or bioceramics [32, 33]. In some cases, rods, screws, and plates made up of stainless steel or titanium are used along with prostheses to provide optimum stabilization. These implants had some major drawbacks, such as improper fitting and complicated postoperative tests like X-rays, and magnetic resonance imaging (MRI). They can cause the injury of neurological structures and blood vessels as well as collapsing of the instrumented spine. In recent years these implants are being manufactured from carbon fiber/polyetheretherketone (PEEK) and carbon fiber/polysulfone. The composite cage has the advantage of having an elastic modulus similar to the bones, which promotes the maximum growth of the bone into the cage, and also, they are radiolucent in nature. For problems associated with intervertebral discs, artificial discs are the best option. Metal balls and injectable silicone elastomers or hydrogels are being used after discectomy for replacing the nucleus [24, 34]. For disc prostheses, composites are proving to be a good alternative. But the major disadvantage is that the artificial discs cannot be used for long-term applications. In a few cases, a stainless steel rod of adjustable nature, is used for stabilizing the curvature. To overcome the corrosive property of steel, a polymer composite rod has been developed. This composite is made up of braided, unidirectional CF and biocompatible epoxy resin.

1.7.2.3 IMPLANTS FOR JOINT REPLACEMENTS

Joints help in the movement of the body and its various parts, and they are lubricated with the help of a synovial fluid that is viscous in nature. All the joints consist of two opposing articular surfaces, which are further covered and protected by a thin articular cartilage layer. Joint osteoarthritis is a very common disease that results in deterioration in joints and sometimes also causes major deformations in the bone and cartilage of joints. Many times, due to severe metabolic disorders, joints get damaged. Artificial joints are being used to replace various joints in the body. Most common joint replacements have been discussed in further sections.

1.7.2.4 KNEE REPLACEMENT

The knee joint has a very complex biomechanical movement and geometry as compared to the hip joint. In recent days, knee replacement has become very common. A typical knee replacement involves the replacement of femoral and tibial components. Normally, Co-Cr, and Ti alloys, are used for the femoral components, and for the tibial part, it is ultra-high molecular weight polyethylene (UHMWPE), supported by a metallic tibial tray. But UHMWPE causes sinking of the implant as a result of cold deformation. To overcome this drawback, UHMWPE reinforced with CF is being used, and this reinforcement also enhances the tensile yield strength, creep resistance, stiffness as well as fatigue strength of UHMWPE [24, 35].

1.7.2.5 HIP REPLACEMENT

Total hip replacement is very common implantation being carried out in the human body [84]. Hip replacements are of many types and can be differentiated based on the changes in the geometry and material of the acetabular cups, femoral stems as well as the method of fixation. A regular hip replacement procedure consists of a cup-type acetabular section and a femoral module fit into it. This femoral module consists of a head that is designed to fit properly in the acetabular cup. The materials used as implants are stainless steel, Co-Cr, and Ti alloys for femoral shaft and neck, Co-Cr alloy or ceramics, such as alumina and zirconia for the head or the ball. To reduce friction and wearing of metals, Charnley has used an acetabular component made up of polytetrafluorethylene (PTFE) to fit into a femoral module made up of stainless steel [24, 37]. The advantages of using PTFE are low friction of coefficient, excellent thermal stability, hydrophobicity, high chemical stability, and the most important one is inertness in the human body. But PTFE, made acetabular cups used as total hip substitute prostheses showed the disadvantage of unfavorable high distortion and wear. Therefore, for load-bearing applications, ultra-high molecular weight polyethylene (UHMWPE) was used instead of PTFE. To improve the quality of UHMWPE with respect to creep resistance, stiffness, and strength, it was reinforced using CF [37, 38]. Recently ceramics such as dense alumina or zirconia ball are being used for manufacturing acetabular cups. The potential advantages of ceramics usage are low coefficient of friction and wear rate, high hardness as well as compressive strength, and also good biological acceptance. The choice of materials suitable for implant design is very limited because of this application, mostly materials with high strength are needed. Compared to metals, polymers depict more elasticity and also offer better modulus and strength. It was found that composite implants made by using carbon fiber/carbon or carbon fiber/polysulfone showed high static and fatigue strengths which promoted faster bone-bonding compared to the conventional implants. Carbon fiber/polyetheretherketone(PEEK) composite stems showed a mechanical behavior that was similar to that of the femur bone [39, 40]. Representations of knee and hip replacement are shown in Figures 1.7 and 1.8, respectively.

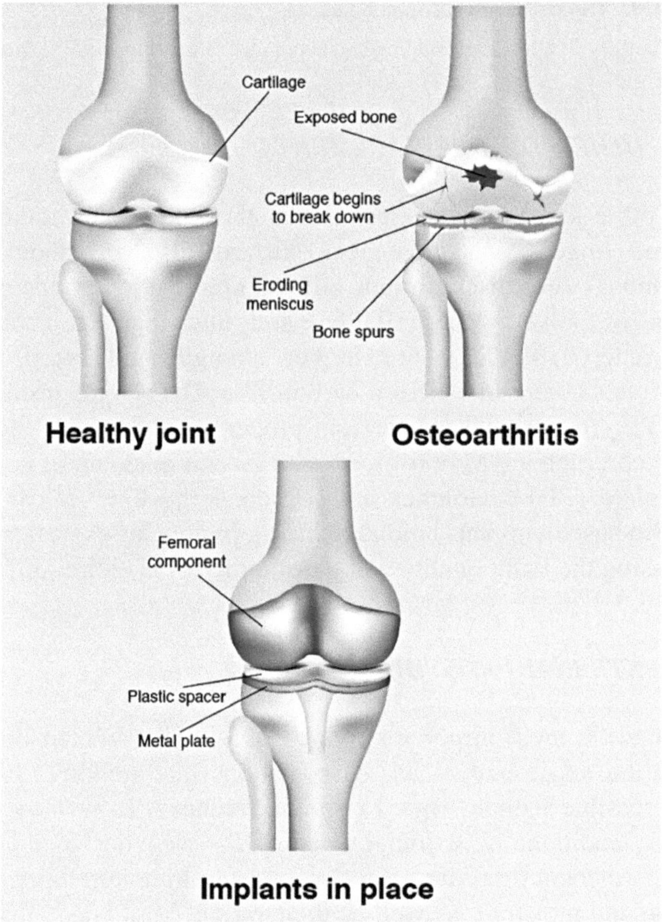

FIGURE 1.7 Knee replacement components.

Source: Courtesy of https://sportssurgeryclinic.com/services/joint-replacement-at-ssc/total-knee-replacement/

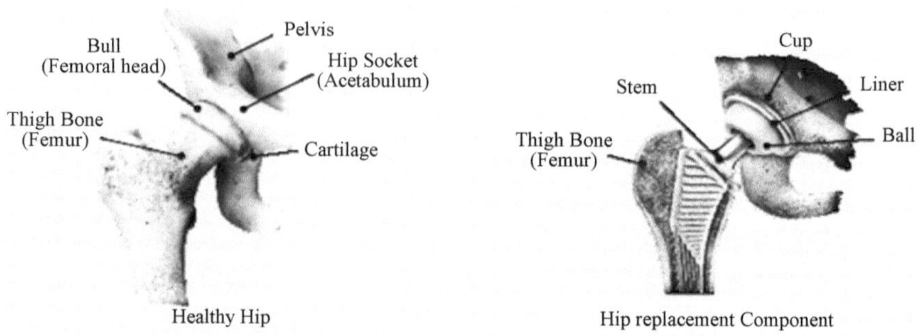

FIGURE 1.8 Hip replacement components.

Source: Courtesy of https://www.jnjmedtech.com/en-US/treatment/hip-replacement

1.7.2.6 OTHER JOINT REPLACEMENTS

Various other joint replacements such as the ankle, toe, shoulder, elbow, wrist, and finger joints have been successfully carried out by using Co-Cr and Ti (cobalt-chromium and titanium) alloys, silicone rubber, high-density polyethylene (HDPE), and ultra-high molecular weight polyethylene(UHMWPE). For higher strength and creep resistance, some implants were made from carbon fiber/UHMWPE instead of only UHMWPE. To improve the flexural properties and tear strength, sometimes silicon rubber was reinforced with fabrics made up of polyethylene terephthalate (PET). Goldner and Urbaniak have revealed that these composite-based implants help in reducing joint pains as well as they help in increasing the functionality and the stability of the joint [36].

1.7.3 EXTERNAL PROSTHETICS

Legs have the most important role to play in the human body that is carrying the entire body load. Also, actions like walking, and running are not possible without legs. Leg bone fractures, as well as injury, are common. Traditionally, artificial wooden legs were designed and used in case of permanent disability. But they were lacking durability, caused by corrosion and moisture, as well as their weight limit was a major hurdle in their application field [24, 41]. To overcome the above disadvantages, fiber-reinforced plastics have gained importance in this area because their usage results in energy conservation. Also, they are safe to handle, have low weight, have good durability, and are available in affordable sizes. Several designs of artificial limbs by using different thermosetting polymer composites are now widely used in these systems. The reinforcing materials used are glass, carbon, or Kevlar fibers [24, 42]. The human leg has three different parts such as socket, shaft, and foot. For transtibial (TT) and transfemoral (TF) prostheses, carbon-fiber-reinforced (CFR) composites have been used. This prosthesis is placed in between the residual limb and shank and helps in managing the load of the body over the ground. But these implants can be subjected to a weight limit of a maximum of 2 kgs only. The matrix used in this composite is composed of rigid and flexible methyl methacrylate (MMA) resin, and its stiffness is matched up to that of the human body socket. In older days, artificial arms made up of SS were being used as a prosthesis, but in recent years, it has been substituted by CFR epoxy tubes. Carbon-fiber woven cloth and epoxy composites are used as an implant for the transfemoral shank. As compared to carbon or nylon composites, the hybrid composites made out of carbon and nylon fibers and having polyester resin as the matrix showed better mechanical properties. The prosthetic material to be used in the foot part, should be capable enough of performing the metabolic activities similar to that of natural foot. The foot prostheses made out of carbon fiber-reinforced epoxy, showed high flexural energy storing capacity along the entire length of the prosthesis. Normally the heel part prostheses are made out of Kevlar and nylon-based composites. The prosthesis was enclosed by using a cover made up of nylon-reinforced silicone elastomer [5]. Artificial sockets are categorized as indirect and direct sockets. The indirect socket is made by wrapping several layers of knitted fabrics, which are impregnated using polyester resin, on the required plastic mold [43]. A direct socket has the benefit of proper fitting between the stump and socket. Direct sockets are made by combining braided or knitted carbon or glass fiber fabrics and water-curable (water-activated) resins. These knitted fabric reinforced sockets are flexible and more comfortable for the patient's stump [24]. In the orthotics field, the traditional cotton fabric and plaster of Paris-made bandages used for supporting injured tissue have been replaced by tailor-made composites in which the matrix is partially cured PU, and the reinforcement is of either fiberglass or polyester knitted fabrics. These laminates show good strength and lower water absorption [24]. An image of leg prostheses is given in Figure 1.9.

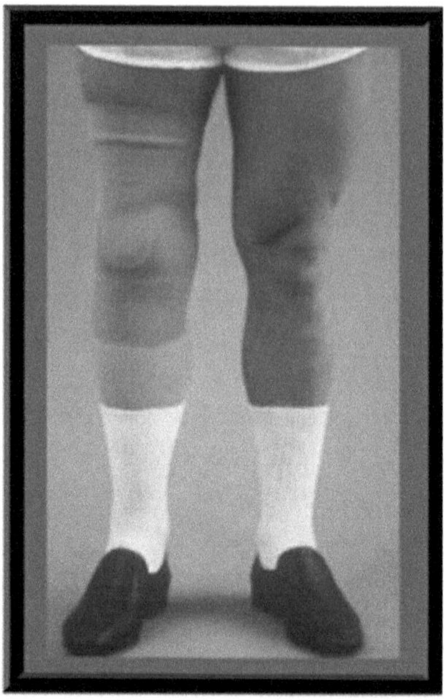

FIGURE 1.9 Leg prostheses.

1.7.4 SOFT TISSUE REPLACEMENT

1.7.4.1 CARDIOVASCULAR GRAFT

Soft tissue replacement and regeneration in vital organs of the human body are very important. In recent years biomaterial suitable for cardiovascular applications has been developed. Normally the cardiovascular tissues are composed of composite materials such as elastin and collagen. Here elastin helps in providing the elasticity, and collagen provides stiffness. Also, the major requirement for the composite to be used here is that it should have mechanical properties equivalent to the original tissue. Millom and Wan [19] have proposed a natural composite made of bacterial cellulose (BC) and polyvinyl alcohol (PVA) to mimic the role of collagen and elastin, respectively, for possible usage as a replacement for a heart valve. Here PVA shows a stress-strain relationship similar to that of human heart tissues. BC has a high elastic modulus, so BC/PVA composite satisfies the property required by cardiovascular tissue and can be used for its replacement [46]. Cardiovascular grafts have also been made by weaving or knitting extruded wall tubes made up of PTFE and PU materials. This composite graft shows anisotropic behavior, similar to the natural artery. Polyethylene terephthalate (PET) (Dacron) made vascular grafts (woven or knitted fabric tubes) are also being successfully used for replacing blood vessels of diameters ranging from 12–38 mm. But this has shown thrombogenic activity and hence is avoided. For replacing blood vessels having a diameter in the range of 6–12 mm, vascular graft has been developed by using PTFE or Gore-Tex. Its mechanical behavior has been found to be more similar to that of host blood vessels compared to non-porous (solid) vascular grafts. Moreover, its inner surface promotes the development of newly formed endothelial tissue lining(neointima) that avoids many complications, such as blood clotting and emboli formation. Researchers have also developed composite grafts using PU (Lycra) fibers as the matrix, and the resin is a mixture of PELA (block copolymer of lactic acid and polyethylene glycol) and PU [48, 49].

1.7.4.2 TENDONS AND LIGAMENTS

Tendons and ligaments are the soft tissue of the human body having load-bearing capacity. The tendon in the human body is first replaced by operation, wherein an implant is placed, which promotes the formation of a new tendon sheath, and in a second operation, the gliding implant is replaced by a tendon graft [24]. Synthetic biopolymers, including polypropylene (PP), PU, ultra-high molecular weight polyethylene (UHMWPE), PTFE, polyethylene terephthalate (PET), Kevlar 49, carbon, and reconstituted collagen fibers in multifilament form or braided form have been used for artificial tendon and ligament synthesis. Few scientists have also developed a ligament prosthesis by reinforcing a hydrogel matrix of poly 2-hydroxyethyl methacrylate (HEMA) with helically wound rigid polyethylene terephthalate (PET) fibers to improve both static and dynamic mechanical behaviors [24, 47].

1.7.5 WOUND DRESSING

Green composites are also being used for dressing injuries/wounds. Different types of chitin and chitosan-based materials are being widely used for these applications. Normally they are used in the form of composites which are further available in the form of gels, nanofibers, films, nonwovens, and scaffolds [50]. Chitin-based composites exhibit excellent biological properties such as biodegradability and biocompatibility and physicochemical properties such as nontoxicity, etc. Bouillat Terrier Orthopedic have developed their first "ecologic orthosis," which is composed of Twinflax 100% flax fabric (235 g/m 2) and an SR Greenpoxy 55 (Sicomin) matrix [51]. More recently, Naseri et al. [52] have developed a chitosan-/polyethylene oxide-composite that has the reinforcement material of chitin nanocrystals (ChNC) suitable for this application.

Montmorillonite (MMT) is a well-known clay that is being used widely for cleaning and protecting the skin, as well as for wound healing. Ul-Islam and co-workers have reported the preparation of BC-MMT nanocomposites through a simple particle impregnation method. This work proved that it is possible to enhance the physicomechanical properties of BC and to create a natural composite which is having extremely large specific surface area and has the ability to form 3D porous structures which are identical to a natural extracellular matrix (ECM). This property increases its demand in the field of wound dressings. Apart from improving the mechanical properties, electrospun composite fibers can also impart additional properties, for example, antibacterial activity by incorporating silver nanoparticles [53] or polypeptides [54]. Son and co-workers [53] have explained the production of electrospun nanocomposite membranes using cellulose acetate/silver nitrate solution. This membrane showed strong antimicrobial activity and was suitably used for wound dressing applications.

KEYWORDS

- **biodegradable**
- **biomedical field**
- **green composites**
- **metal matrix composites**
- **natural fibers**
- **polymer is polylactic acid**
- **renewable sources**

REFERENCES

1. Bhaskar, C. K., Rakesh, C., & Vishal, K., (2015). Recent advances in green composites: A review. *International Journal for Technological Research in Engineering (IJTRE), 2*(7), March-
2. Ibrahim, I. D., Sadiku, E. R., Jamiru, T., Hamam, A., & Kupolati, W. K., (2017). Applications of polymers in the biomedical field. *Current Trends in Biomedical Engineering & Biosciences.*
3. Zunjarrao, B. K. (2013). *Green Composites.*
4. Chitra, N. J., (2015). *Chapter I: Introduction, Studies on Development and Characterization of Polypropylene Based Biocomposites.*
5. Arif, I., (2004). Chapter 12: Biomedical composites. *Standard Handbook of Biomedical Engineering and Design.*
6. Velmurugan, R. (2019). *Composite Materials.* www.slideshare.net/kamblezunjar/green-composites-1-1-2, Department of Aerospace Engineering, IIT Madras, Module 1, NPTEL Lecture.
7. Anil, N. N., (2010). *'Green' Composites: Where We Are and Where We Are Headed.* Midwest biopolymers and biocomposites workshop Ames, Iowa.
8. Ibrahim, I. D., Sadiku, E. R., Jamiru, T., Hamam, A., & Kupolati, W. K., (2017). Applications of polymers in the biomedical field, *Current Trends in Biomedical Engineering & Biosciences, 4*(5). CTBEB.MS.ID.5555650.
9. Bassas-Galia, M., Follonier, S., Pusnik, M., & Zinn, M., (2016). 2-natural polymers: A source of inspiration. In: Perale, G., & Hilborn, J., (eds.), *Bioresorbable Polymers for Biomedical Applications* (pp. 31–64).
10. Mao, C., Liang, C., Luo, W., Bao, J., Shen, J., et al., (2009). Preparation of lotus-leaf-like polystyrene micro-and nanostructure films and its blood compatibility. *Journal of Materials Chemistry, 19*(47), 9025–9029.
11. Koronis, G., Arlindo, S., & Samuel, F., (2016). Chapter 10: Applications of green composite materials. *Biodegradable Green Composites* (1st edn.).
12. Gourav, G., Ankur, K., Rahul, T., & Sachin, K., (2016). Application and future of composite materials: A review. *International Journal of Innovative Research in Science, Engineering and Technology, 5*(5).
13. Muzzarelli, R. A. A., (1983). Chitin and its derivatives: New trends of applied research. *Carbohydrate Polymers, 3*, 53–75.
14. Guitian, O. N., Sirgado, T., Reis, L., Pinto, L. F. V., Da Silva, C. L., Ferreira, F. C., & Rodrigues, A., (2014). In vitro assessment of three dimensional dense chitosan-based structures to be used as bioabsorbable implants. *Journal of the Mechanical Behavior of Biomedical Materials, 40*, 413–425.
15. Zubir, N., & Pushpanathan, K., (2016). Silk in biomedical engineering: A review. *International Journal of Engineering Inventions, 5*(8), 18–29.
16. Wang, Z., Hu, Q., & Cai, L., (2010). Chitin fiber and chitosan 3D composite rods. *International Journal of Polymer Science, 2010*, 7.
17. Oh, D. X., & Hwang, D. S., (2013). A biomimetic chitosan composite with improved mechanical properties in wet conditions. *Biotechnology Progress, 29*, 505–512.

18. Ehrlich, H., Simon, P., Motylenko, M., Wysokowski, M., Bazhenov, V. V., Galli, R., Stelling, A. L., et al., (2013). Extreme biomimetics: Formation of zirconium dioxide nanophase using chitinous scaffolds under hydrothermal conditions. *Journal of Materials Chemistry,* 5092–5099.
19. Hoi-Yan, C., Mei-po, H., Kin-Tak, L., Francisco, C., & David, H., (2009). Natural fiber-reinforced composites for bioengineering and environmental engineering applications. *Composites: Part B, 40*, 655–663.
20. Meinel, L., Hofmann, S., Karageorgiou, V., Zichner, L., Langer, R., Kaplan, D., et al., (2004). Engineering cartilage-like tissue using human mesenchymal stem cells and silk protein scaffolds. *Biotechnol. Bioeng., 88*, 379–391.
21. www.cmu.edu (Carnegie Mellon University, Biomedical Engineering Department). Available from: http://www.cs.cmu.edu/people/tissue/tutorial.html (accessed on 7 June 2022).
22. Seal, B., & Panitch, A., (2001). Polymeric biomaterials for tissue and organ regeneration. *Materials Science and Engineering [R], 262*, 1ñ84.
23. Lohr, J. M., James, K. V., Hearn, A. T., & Ogden, S. A., (1996). ìLessons learned from the DIASTAT vascular access grafting. *American Journal of Surgery, 172*, 205ñ209.
24. Arijit, N., Tiyasha, K., & Chiranjib, B. (2016). *Application of Bioactive Composite Green Polymer for the Development of Artificial Organs* (Vol. 5). Materials Research Foundations.
25. Black, J., (1988). *Orthopaedic Biomaterials in Research and Practice*. Churchill Livingstone. New York.
26. Chao, E. Y. S., & Aro, H. T., (1997). Biomechanics of fracture fixation. In: Mow, V. C., & Hayes, W. C., (eds.), *Basic Orthopaedic Biomechanics* (pp. 317–350). Lippincott-Raven, Philadelphia.
27. Roman, J. S., & Garcia, P. G., (1991). Partially biodegradable polyacrylic-polyester composites for internal bone fracture fixation. *Biomaterials, 12*, 236–241.
28. Rubin, L. R., (1983). Biomaterials in maxillo-facial surgery. In: Szycher, M., (ed.), *Biocompatible Polymers, Metals, and Composites* (pp. 941–951). Lancaster, USA: Technomic Publishing.
29. Ignatius, A., Unterricker, K., Wenger, K., Richter, M., & Claes, L., (1997). A new composite made of polyurethane and glass-ceramic in a loaded implant model: A biomechanical and histological analysis. *Journal of Material Science Materials in Medicine, 8*(12), 753–756.
30. Andersson, G. B. J., (1993). Intervertebral disk. In: Wright, V., & Radin, E. L., (eds.), *Mechanics of Human Joints: Physiology, Pathophysiology, and Treatment* (pp. 293–311). Marcel Dekker Inc. New York.
31. Marcolongo, M., Ducheyne, P., Garino, J., & Schepers, E., (1998). Bioactive glass fiber/polymeric composites bond to bone tissue. *Journal of Biomedical Material Research, 39*(1), 161–170.
32. Valdevit, A. D. C., Inoue, N., Mac Williams, B. A., & Anderson, L. L., (1996). Methods for mechanical testing of spinal constructs. *Spine, 10*(2), 231–248.
33. Brantigan, J. W., Steffee, A. D., & Geiger, J. M., (1991). A carbon fiber implant to aid interbody lumbar fusion mechanical testing. *Spine (Phila Pa 1976), 16(6 Suppl)*, S277–282.

34. Bao, Q. B., McCullen, G. M., Higham, P. A., Dumbleton, J. H., & Yuan, H. A., (1997). The artificial disc: Theory, design and materials. *Neurosurgery, 41*(5), 1203–1206.
35. Silverton, C., Rosenberg, A. O., Barden, R. M., Sheinkop, M. B., & Galante, J. O., (1996). The prosthesis-bone interface adjacent to tibial components inserted without cement. *Journal of Bone and Joint Surgery American, 78*(3), 340–347.
36. Goldner, J. L., & Urbaniak, J. R., (1973). The clinical experience with silicone-Dacron metacarpophalangeal and interphalangeal joint prostheses. *Journal of Biomedical Material Research, 7*(3), 137–163.
37. Sclippa, E., & Piekarski, K., (1973). Carbon fiber reinforced polyethylene for possible orthopedic uses. *Journal of Biomedical Material Research, 7*(1), 59–70.
38. John, K. R., (1983). Applications of advanced composites in orthopedic implants. In: Szycher, M., (eds.), *Biocompatible Polymers, Metals, and Composites* (pp. 861–871). Lancaster: Technomic Publishing.
39. Peter, T., Tognini, R., Mayer, J., & Wintermantel, E., (1997). Homoelastic, anisotropic osteosynthesis system by net-shape processing of endless carbon fiber-reinforced polyetheretherketone (PEEK). In: Goh, J. C. H., & Nather, A., (eds.), *Proc. of 9th Conference on Biomedical* Engineering (pp. 317–319). Singapore, National University of Singapore.
40. Williams, D. F., (1988). Consensus and definitions in biomaterials. In: De Putter, C. C., De Lange, K., De Groot, K., & Lee, A. J. C., (eds.), *Advances in Biomaterials* (pp. 11–16). Elsevier Science, Amsterdam.
41. Robin, G. C., (1981). Below-knee drop-foot braces: Stresses during use and evaluation of design. In: Ghista, D. N., (ed.), *Biomechanics of Medical Devices* (pp. 535–567). Marcel Dekker Inc., New York.
42. Coombes, A. G. A., Greenwood, C. D., & Shorter, J. J., (1996). Plastic materials for external prostheses and orthoses. In: Wise, D. L., Trantolo, D. J., Altobelli, D. E., Yaszemski, M. J., & Gresser, J. D., (eds.), *Human Biomaterials Applications* (pp. 215–255). Humana Press Totowa, New York.
43. Tallent, M. A., Cordova, C. W., Cordova, D. S., & Donnelly, D. S., (1990). Thermoplastic fibers for composite reinforcement. In: Lee, S. M., (eds.), *International Encyclopedia of Composites* (pp. 466–480). VCH Publishers, New York.
44. Ramakrishna, S., Mayer, J., Wintermantel, E., & Kam, W. L., (2001). Biomedical applications of polymer-composite materials: A review. *Composites Science and Technology, 61*(9), 1189–1224.
45. Salernitano, E., & Migliaresi, C., (2003). Composite materials for biomedical applications: A review. *Journal of Applied Biomaterials & Biomechanics,* 3–18.
46. Ana, B., Isabel, F., & João, P. B. (2013). Chapter 3: Cellulose-based composite systems for biomedical applications. *Biomass-based Biocomposites.*
47. Iannace, S., Sabatini, G., Ambrosio, L., & Nicolais, L., (1995). Mechanical behavior of composite artificial tendons and ligaments. *Biomaterials, 16*(9), 675–680.
48. Gershon, B., Cohn, D., & Marom, G., (1992). Compliance and ultimate strength of composite arterial prostheses. *Biomaterials, 13*(1), 38–43.
49. Gershon, B., Cohn, D., & Marom, G., (1990). Utilization of composite laminate theory in the design of synthetic soft tissues for biomedical prostheses. *Biomaterials, 11*(8), 548–552.

50. Muzzarelli, R. A. A., (2012). Nanochitins and nanochitosans, paving the way to eco-friendly and energy-saving exploitation of marine resources. In: Hoefer, R., (ed.), *Polymer Science: A Comprehensive Reference* (pp. 153–164). Elsevier, Amsterdam.
51. Anonymous, (2012). Advanced composite solutions (Vol. 72, pp. 15–19). JEC Composite.
52. Naseri, N., Algan, C., Jacobs, V., John, M., Oksman, K., & Mathew, A. P., (2014). Electrospun chitosan-based nanocomposite mats reinforced with chitin nanocrystals for wound dressing. *Carbohydrate Polymers, 109*, 7–15.
53. Son, W. K., Youk, J. H., & Park, W. H., (2006). *Carbohydrate Polymers, 65*, 430.
54. Miao, J., Pangule, R. C., Paskaleva, E. E., Hwang, E. E., Kane, R. S., Linhardt, R. J., & Dordick, J. S., (2011). *Biomaterials, 32*, 9557.
55. Sanjay, M.R., Arpitha, G.R., Naik, L.L., Gopalakrishna, K., Yogesha, B. Applications of Natural Fibers and Its Composites: An Overview. Natural Resources, Vol.7 No.3, 2016.
56. Mushtaq, S., and Majeed, M.I. Editorial: Biopolymers for Medical Applications. 2022 Technology Times. https://technologytimes.pk/2017/08/15/biopolymers-for-medical-applications/

CHAPTER 2

Characterization of Natural Particulates Filled and E-Glass Fiber-Reinforced Sandwich Polymer Composites

C. BALAJI AYYANAR[1] and K. MARIMUTHU[2]

[1]*Assistant Professor, Department of Mechanical Engineering, Coimbatore Institute of Technology, Coimbatore–641014, Tamil Nadu, India, E-mail: balajiayyanar2007@gmail.com*

[2]*Professor, Department of Mechanical Engineering, Coimbatore Institute of Technology, Coimbatore–641014, Tamil Nadu, India*

ABSTRACT

The 10, 20, 30, 40, and 50 wt.% of (i) fish scale particulate filled epoxy composite, and (ii) fish scale particulate filled and E-glass fiber laminate reinforced (Woven Roving WR 400 GSM) sandwich composites were fabricated with different weight percentage. The two different combinations of composite with different wt.% of reinforcement were evaluated the mechanical characteristics such as tensile, compression, hardness, flexural strength, fatigue strength, and surface morphologies (FESEM: field emission scanning electron microscope), and EDX analysis was carried out, and results were compared. The maximum tensile strength of 65 MPa (megapascal), compressive strength of 153 MPa, shore D hardness of 99 SHN, and flexural strength of 355 MPa were found. By incorporating fiber laminate on both sides of the sandwich composite compared with natural particulates-filled epoxy composites, the mechanical properties have been enhanced to some extent.

Advances in Diverse Applications of Polymer Composites: Synthesis, Application, and Characterization.
Suji Mary Zachariah, Yang Weimin, Maciej Jaroszewski, & Sabu Thomas (Eds.)

2.1 INTRODUCTION

The recent years, the light core sandwich structure has been considered a substitute for many conventional structures. A sandwich laminated composite grouping of two or three different materials that are bound to each other so as to utilize the properties of each separate reinforcement. Improvement of lightweight core and composite materials effort to reduce the weight of sandwich composites and to increase their strength and stiffness. Fiber-reinforced sandwich composites are being increasingly brought into play in applications requiring both high strengths with low weight [1].

Recently hydroxyapatite has been introduced as an artificial bone graft material in different medical domains appropriate to their similar chemical composition of bones [2]. The calcium deficient hydroxyapatite mineral phase found in the fish scale (Pagrus) has a higher tensile strength of 90 MPa, and demineralized scales have a lower tensile strength of (36 MPa), which depends on the interactions between the apatite and collagen fibers [3]. Fish scales (Lobea rohita) reinforced with epoxy composites indicated a tensile strength of 5 wt.% (66 MPa), 10 wt.% (64 MPa), and 15 wt.% of which (61 MPa) were estimated with neat epoxy that is about 70 MPa [2]. Scale (Megalops Atlanticus) consists of collagen-reinforced apatite. The tensile strength of scales was found in the head, middles, and tail regions in the possible three primary orientations 0° (24 MPa), 45° (20 MPa), and 90° (15 MPa), respectively [4]. The hydroxyapatite was extracted from pork bones through hydrolysis of lactic acid and pre-calcination at 600°C at the calcination stage (750–950°C) [4]. It can also be extracted from scales (tilapia) by calcination and acid-base treatment [5]. It was extracted from fish scale by enzymatic hydrolysis and has increased MG 63 growth which is a capable biomaterial for non-natural bone fabrication [6].

Scales have both flexibility and protection. Since tissues form a natural, flexible armor that guards essential tissue and vital organs [7]. The FTIR spectrum of scale and extorted collagen shows amide I (1,640 cm^{-1}), amide II (1,534 cm^{1}), amide III (1,226 cm^{-1}), visible bands in scales at 1,073 cm^{-1}, 600 cm^{-1}, and 545 cm^{-1} shows the asymmetric stretching vibrations symmetric bending vibration, and asymmetric bending vibration of a phosphate group (PO_4^{3-}) [8]. The composite consists of the probable debonding of the outside facings of the sandwich (skins), which must possess considerable rigidity and strength of the center of the sandwich (core), which is required to acquire a low weight and enough shear stiffness. The numerical simulation and characterization of a foam/glass fiber sandwich composite are considered a lightweight material for various engineering applications [9]. It was chosen as fillers in HDPE, as polyethylene (PE) thermoplastic has excellent biocompatibility and mechanical properties [10].

The objective of the study is to develop composite materials by using naturally held hydroxyapatite fish scale particulates. The composite was fabricated with different weight percentage fish scale reinforcement without E glass fiber. The sandwich epoxy composite was prepared with different wt.% fish scale particulates. The single layer of E glass fiber was placed outside of the composite.

2.2 EXPERIMENTAL DETAILS

2.2.1 MATERIALS

The fish scales are (Labeo L. Catla) collected from a fish market in the local area of Omalur, Salem, Tamil Nadu (India). The epoxy resin (LY 556), the hardener HY-951, and E Glass fiber Woven Roving WR 400 GSM used for this experiment were procured from Covai and Seenu brothers, Coimbatore, Tamil Nadu (India). The properties of the fish scale, Epoxy, and E Glass fiber Woven Roving WR 400 GSM were given in Tables 2.1–2.3.

TABLE 2.1 Properties of Fish Scale

Appearance	White Color
Density (g/cm^3)	0.9
Particulates size (μm)	20–30
Genus	Labeo
Species	L. Catla

2.2.2 FABRICATION OF COMPOSITE

The fully grown fish scales are washed to take out dust and impurities from their surface. Cleaned scales are kept in the sunshine for one day, and dried fish scales are then ground into powder in a flour mill of 20–30 microns. These biodegradable scale particulates were used as the reinforcing phase with different weight percentages (i) 10, (ii) 20, (iii) 30, (iv) 40, and (v) 50 wt.%). (i) fish scale particulates filled composite is shown in Figure 2.1; and (ii) fish scale particulate filled, and e-glass fiber laminate reinforced (Woven Roving WR 400 GSM) sandwich composites were fabricated is shown Figure 2.2.

TABLE 2.2 Properties of Epoxy Resin

Density (g/cm3)	1.1
Compressive strength (N/mm2)	190
Glass transition temperature (T_g)	120–130°C
Tensile strength (N/mm2)	85
Young's modulus (N/mm2)	10,500
Flexural strength(N/mm2)	112
Flexural modulus (N/mm2)	10,000
Elongation at break (%)	50%

TABLE 2.3 Properties of Woven Roving's E-Glass Fiber

Density (g/cm3)	2.54

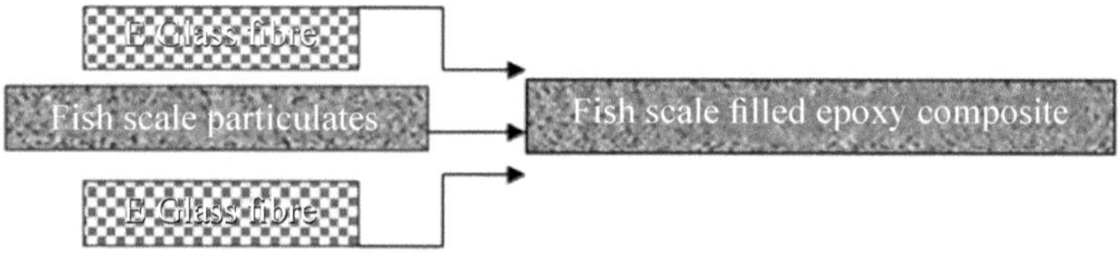

FIGURE 2.1 Schematic view of epoxy composites.

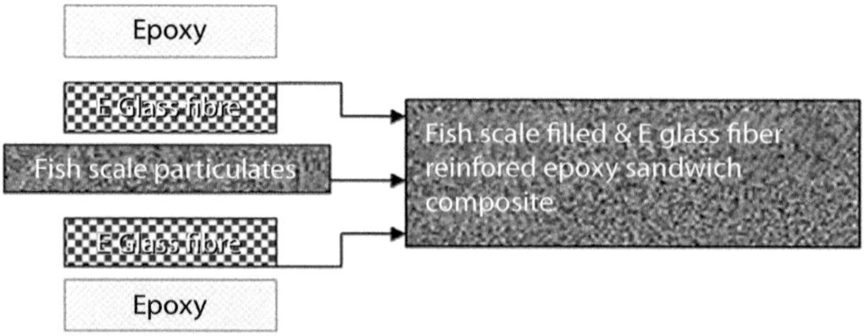

FIGURE 2.2 Schematic view of sandwich composites.

2.2.3 SANDWICH COMPOSITE FABRICATION

The sandwich composites were fabricated by using low-temperature cured matrix epoxy resin, and the corresponding hardener is mixed in proportions of 10:1 by weight percentage as suggested. The schematic outlook of fish scale particulates with different weight percentages reinforced epoxy composite is exposed in Figure 2.3.

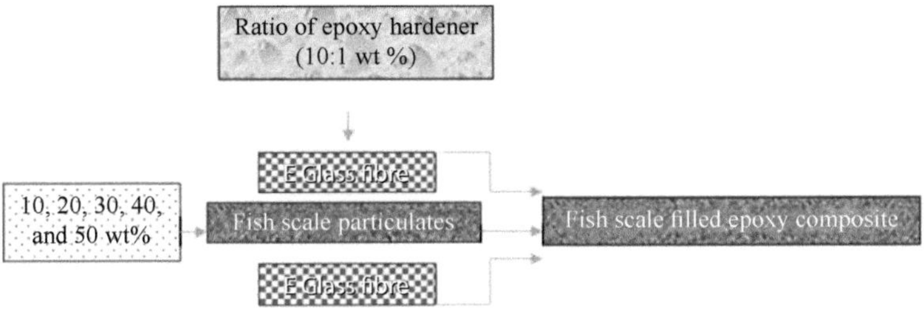

FIGURE 2.3 Schematic view of fish particulates filled.

The schematic outlook of fish scale particulates are different weight percentages were filled, and E glass fiber mat reinforced with the outside of the epoxy composite is shown in Figure 2.4. The matrix and the natural particulates are thoroughly mixed in the chamber before reaching the pot time and poured into the size the block of 300 mm, 150 mm, and 3 mm. The mixture is stirred by hand in the chamber until the particulates are dispersed homogeneously and poured into the mold. Each specimen have molded with reinforcement of different weight parentage such as: (i) 10; (ii) 20; (iii) 30; (iv) 40; and (v) 50 wt.%. The casted specimens of the required size of dimension were prepared using a suitable cutter for physical, mechanical characterization, and microstructure investigation. Figure 2.3 shows the mold die is made of steel which has top and bottom die replicas of blocks of size (length 300 mm, width 150 mm, and thickness 3 mm) in the formed cavity with suitable allowance to meet the required specifications. Fabrication was made by using an open type of compression molding technique. The upper die of the mold was closed and pressed at a pressure of 0.25 Kg/cm^2 at room temperature and cured. Firstly, an epoxy composite was prepared with different weight percentage fish scale reinforcement (10, 20, 30, 40, and 50 wt.%) without fiber laminates. Secondly, the sandwich epoxy composite was fabricated with different wt.% ((i) 10; (ii) 20; (iii) 30; (iv) 40; and (v) 50 wt.%) fish scale particulates. The layer of E glass fiber was placed bottom and top of the composite.

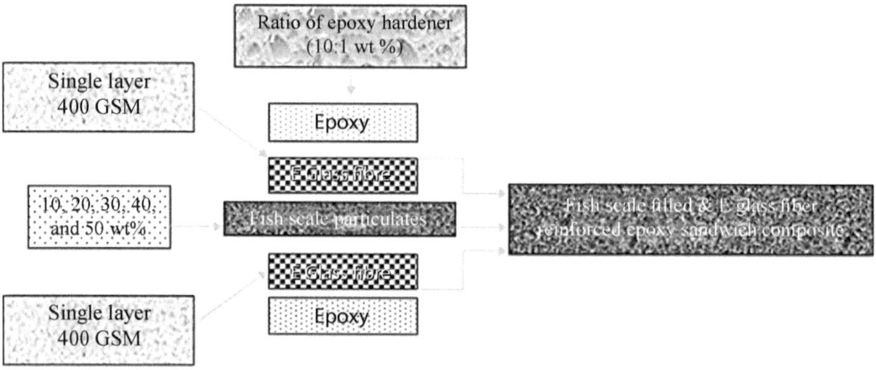

FIGURE 2.4 Schematic view of fish scale particulates filled and e glass fiber.

2.2.4 MECHANICAL TESTING

The tensile testing was performed as per ASTM D638 under ambient conditions on a Deepak Poly Plast (Gujarat, India) universal tester at a speed of 2 mm/min. The specimens (span length 90 mm, gauge length 40 mm, thickness 3 mm, and width 7 mm) were prepared using a plastic cutter. The same compositions of six specimens were tested, and the average values were reported. The compression testing was performed under ambient conditions on a Deepak Poly Plast (Gujarat, India) universal tester at a speed of 2 mm/min according to ASTM D695. The specimens (span length 26 mm, width, and thickness 12.7 mm) were prepared using a plastic cutter. The same compositions of six samples were tested, and the average values were reported.

The Shore D Hardness testing was performed under ambient conditions on a Shore 'D' machine at the Scientific and Industrial Testing and Research Center. The specimens (length 30 mm, and width 10) were prepared using a plastic cutter. The same compositions of six samples were tested, and the average values were reported. The flexural testing was performed as per ASTM D790 under ambient conditions on a Deepak Poly Plast (Gujarat, India) universal tester at a speed of 2 mm/min. The specimens (span length 80 mm, width 13 mm, and thickness 3 mm) were prepared using a plastic cutter. The same compositions of six samples were tested, and the average values were reported. The microstructure and EDX analysis of fish scale particulates filled epoxy composite specimens were examined using a ZEISS Sigma 300 scanning electron microscope (SEM; Bangkok, Thailand). The sample was gold-sputtered before examining the specimen. The FTIR analysis of fish scale particulates was examined. The spectrometer reveals the presence of vibration of the groups in the sample from 500–5,000 cm^{-1}.

2.3 RESULTS AND DISCUSSION

2.3.1 TENSILE BEHAVIOR OF FISH SCALE/SANDWICH COMPOSITES

The natural fiber fish scale particulates reinforced epoxy composites i) 10 wt.%, ii) 20 wt.%, iii) 30 wt.%, iv) 40 wt.%, and v) 50 wt.% specimens, and fish scale particulates filled and reinforced with single laminate E glass Woven Roving WR 400 GSM fiber epoxy composites with a different weight percentage of (i) 10 wt.%, (ii) 20 wt.%, (iii) 30 wt.%, (iv) 40 wt.% and (v) 50 wt.% specimens were carried out tensile strength as per the standard ASTM 638. The graph was plotted, and the tensile strength of both the different composites was compared, as exposed in Figure 2.5. The tensile strength was high at 10 wt.% and it was gradually decreasing by varying the particulate 20 wt.%, 30 wt.%, 40 wt.%, and 50 wt.%, respectively. The tensile strength of 10 (wt.%) without and with E glass fiber laminate composite was increased from 29 MPa to 65 MPa. The strength was gradually decreased by increasing the particulates content, which was observed in the test. The composites do not transfer the load between the matrix and reinforcement due to the unavoidable few bubbles and voids present in the specimens. Also, dropping the tensile strengths of the composites so considerably due to the least bonding between reinforcement and matrix. It was improved by incorporating the single layer of E glass Woven Roving (WR) 400 (gram per square meter) GSM laminate at the bottom and top of the fish scale particulates-filled sandwich composites. The strength has been improved for fish scale particulate epoxy composite by incorporating E glass fiber in the bottom and top of the specimens, and its properties have been studied, and the results were compared. The particulates-filled polymer composites are not sufficient to carry the tensile load large hence it fails at below reaching 30 MPa with increasing wt.% reinforcement; further, it reduces its tensile strength, which is shown in Figure 2.6.

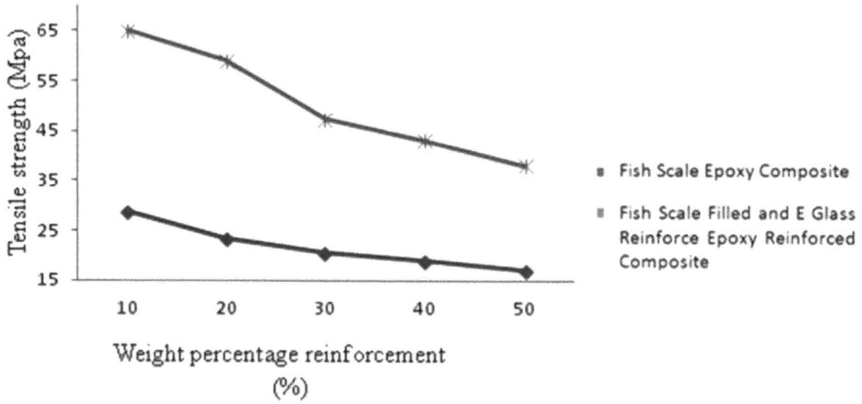

FIGURE 2.5 Tensile strength (MPa) of epoxy composites with different weight percentage reinforcement.

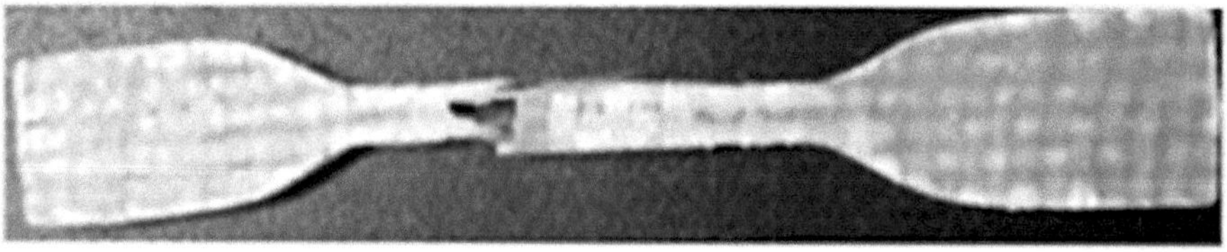

FIGURE 2.6 Tensile failure of composites.

The tensile strength of the composites has been improved from 29 MPa to 65 MPa, respectively, for the same (10 wt.%) reinforcement composites without E and with glass laminate. By incorporating E-Glass fiber laminate, the tensile strength has been improved. The sandwich composite can withstand more force (N) than the composites without E-Glass fiber laminate is exposed in Figure 2.7. The fiber could carry more force and withstanding high load carrying capacity, which has been revealed in microstructure studies through FESEM.

The fractured surfaces of tensile test specimens are examined in the SEM, which shows the fractured region in brittle nature without yielding is marked and shown in the microstructure image in Figure 2.8.

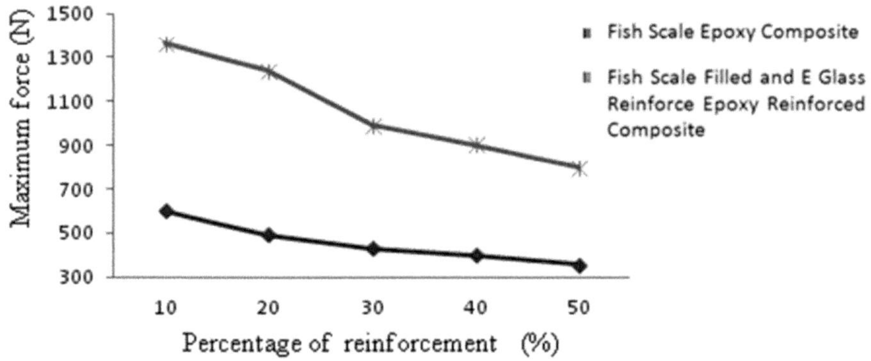

FIGURE 2.7 Tensile forces with a different weight percentage of reinforcement.

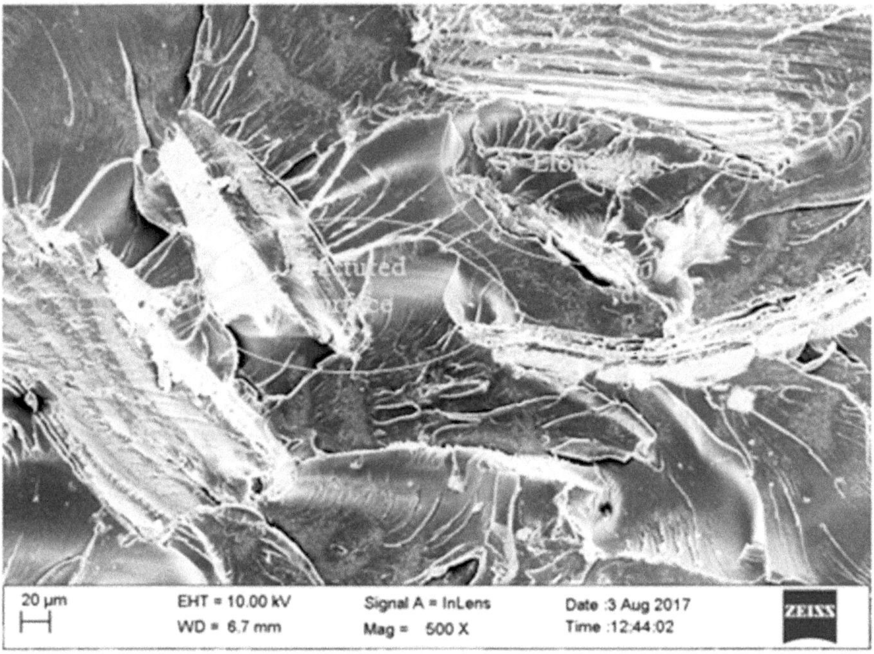

FIGURE 2.8 Morphology of tensile failure surface of epoxy composites without E glass laminate.

The particulates are dispersed in the matrix, which is examined in field emission scanning using an electron microscope of tensile failure composites, as shown in Figures 2.9 and 2.10. The microstructure study

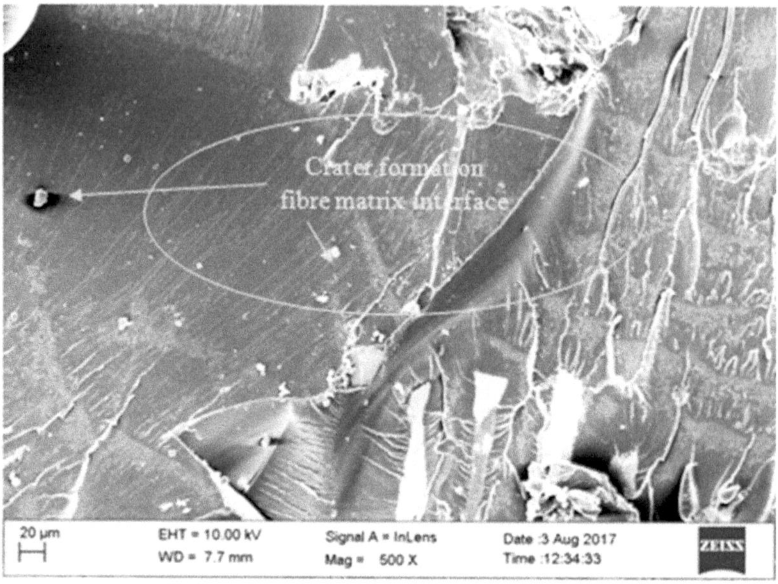

FIGURE 2.9 Morphology of particulates dispersion of epoxy composites without E glass fiber laminates.

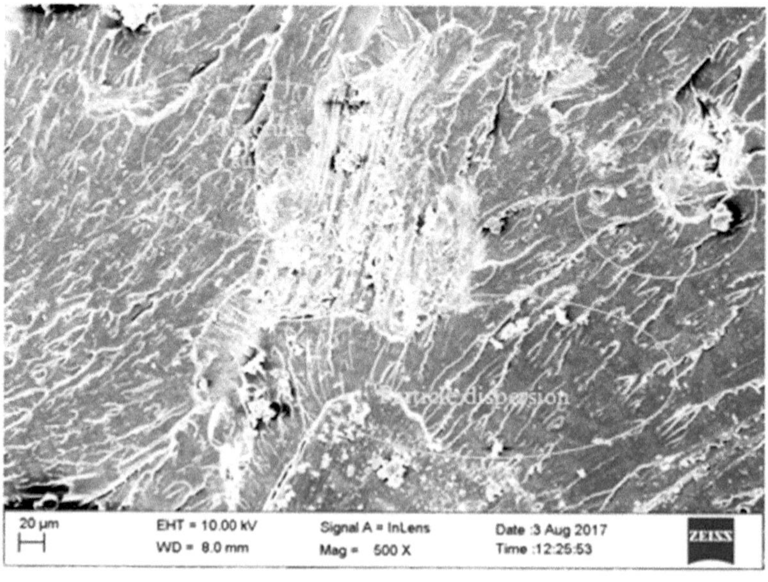

FIGURE 2.10 Morphology of fish scale dispersion of epoxy composites without E glass fiber laminate.

has revealed that the single layer E glass fiber in both sides of sandwich composite has enhanced the strength better than fish scale composites. It can be further increased by incorporating a greater number of layers. The fiber has been deboned and fractured during the tensile strength test when the load increases, which was revealed in FESEM as exposed in Figure 2.11.

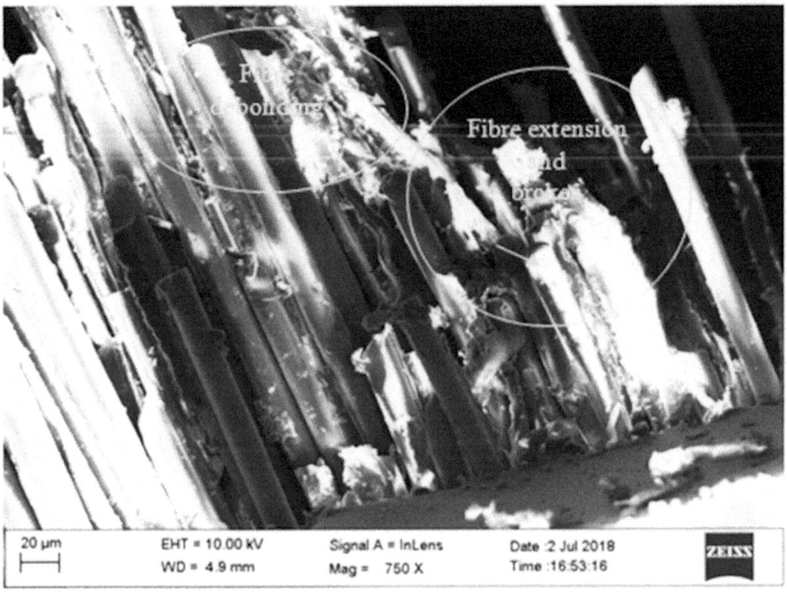

FIGURE 2.11 Morphology of fractured tensile surface of sandwich epoxy composites with E glass fiber laminate.

The E glass fiber interface in fish scale particulates has been revealed in microstructure studies, as shown in Figures 2.12 and 2.13.

2.3.2 COMPRESSION TEST

The compression test is carried out as per the ASTM D 695 standard for each specimen and plotted on the graph for different weight percentages of reinforcement. The compressive strength of fish scale particulate composite gradually increases.

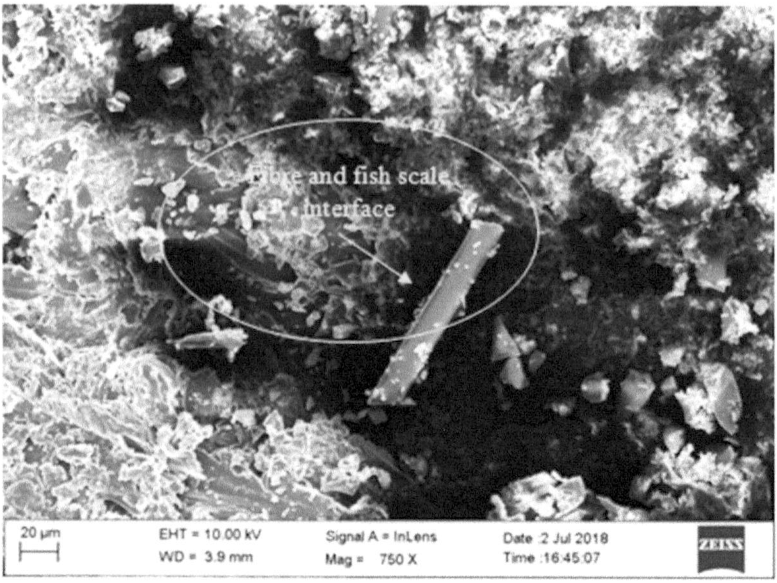

FIGURE 2.12 Morphology of E glass fiber laminate interface in fish scale particulate epoxy composite.

FIGURE 2.13 Morphology of E glass fiber laminate interface in fish scale particulate epoxy composite.

The results revealed that the compressive strength had been improved from 148 Mpa to 153 Mpa, respectively, for the same 50 wt.% of fish scale particulates reinforced composites without E glass and with E glass laminate. The composites which have 50 wt.% particulate carrying maximum compressive strength is about 153 Mpa as shown in Figure 2.14. The other reason is the amount of reinforcement is nearly equivalent to the matrix so that particulates are adjoining together and there is poor load transferring between matrix and reinforcement.

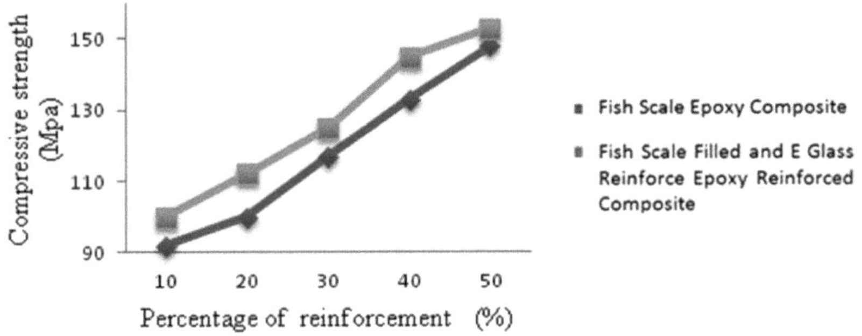

FIGURE 2.14 Variation of compressive strength (MPa) of with and without E glass fiber laminate and composites.

2.3.3 HARDNESS TEST

The hardness test is carried out as per the ASTM D2240 standard for each specimen. Shore D hardness tester is used for thermosetting polymer, force applied to the specimen is 4.55 kg, and 30° cone indenter is used for the testing. The hardness values are measured at five different positions at each sample, and the graph is plotted. The Shore D hardness of the epoxy composites was increased by increasing the reinforcement, as shown in Figure 2.15. The highest hardness value was found to be 99 SHN of 50 wt.% reinforcement of composite. The matrix and reinforcement are touched together, and an interface can transfer loads more effectively, which enhances hardness.

2.3.4 FLEXURAL TEST

The 3-point flexural strength of the epoxy composite increases up to 30 wt.%; after that, it gets decreases. The decline in the flexural strength of these scale particulates reinforced epoxy composites due to the interface strength between the reinforcement and matrix may be too deprived to transfer the strength. Also, corner points in the regular and irregular particulates result in stress concentration.

$$\sigma_f = \frac{3Fl^2}{2bd^2}$$

where; F is maximum load; (b, d, and l) width, thickness, and span length of the specimen.

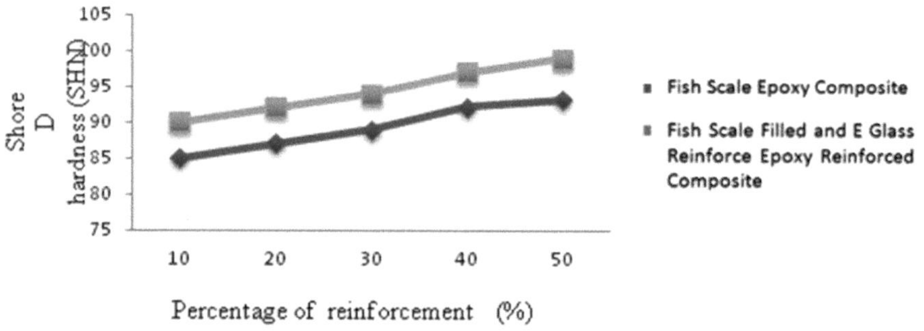

FIGURE 2.15 Variation of the hardness of composites with different particulate content.

2.3.5 *FATIGUE TEST*

The fatigue test is carried out as per the ASTM D606 standard for each specimen in tension-tension mode and found the number of working cycles without failure by varying the stress below the ultimate stress is shown in Figure 2.17. The samples are fractured immediately when the stress level is increased by more than 50% ultimate strength. It is inferred that all the samples can withstand more than 25,000 cycles by varying stress below 50% ultimate strength (Figures 2.16 and 2.17).

2.3.6 *EDX ANALYSIS*

The EDX analysis of scale particulates reinforced epoxy composites gave an organic and inorganic composition of C, O, N, Ca, P, Cl, and S were obtained as 67.36, 24.75, 4.20, 2.55, 0.98, 0.10, and 0.06 wt.% as depicts in Figure 2.18.

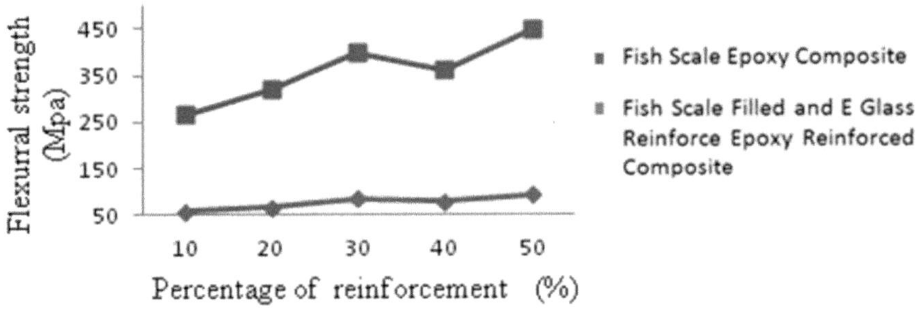

FIGURE 2.16 Variation of flexural strength of composites with different particulate content.

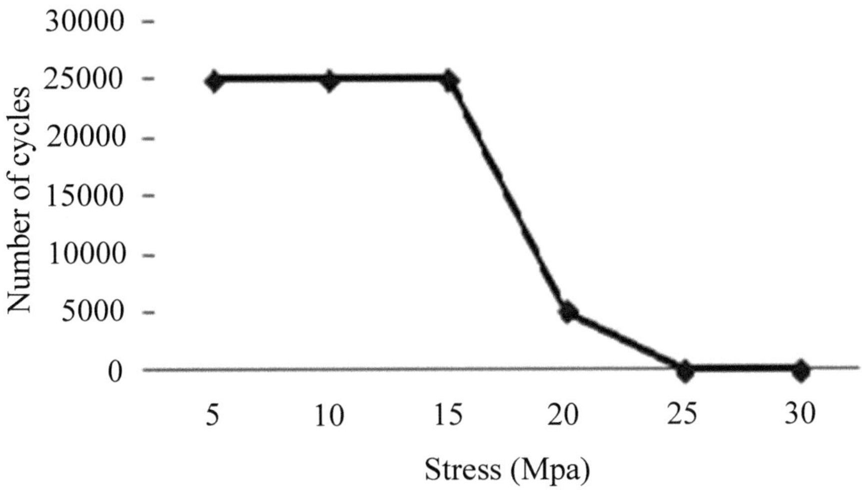

FIGURE 2.17 Variation of the hardness of composites with different particulate content.

2.3.7 FTIR ANALYSIS

Figure 2.19 shows the FTIR spectrum of fish scale particulates ranges from 500–5,000 cm^{-1}. The characteristic band around 3,743 cm^{-1} is observed; it shows the presence of the OH group. The bands around 3,273 cm^{-1}, depict the amide group is most prominent in the fish scale. Apart from that, the band at 1,642 cm^{-1} confirms the existence of bending vibration of amide I in the fish scale. The amide II has a vibration band around 1,532 cm^{-1}, also present in the scale. The band at 1,030 cm^{-1} consigned to amide III, which is also revealed in the fish scale. Moreover, the week bands at 1,030 cm−1 and 606 cm−1 indicate the vibrations of a phosphate group (PO_4^{3-}) on the fish scale.

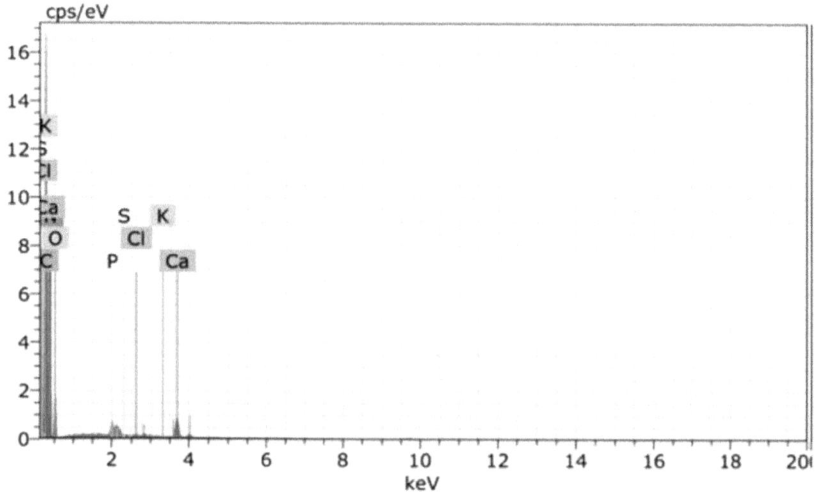

FIGURE 2.18 EDX analysis of fish scale reinforced epoxy composites.

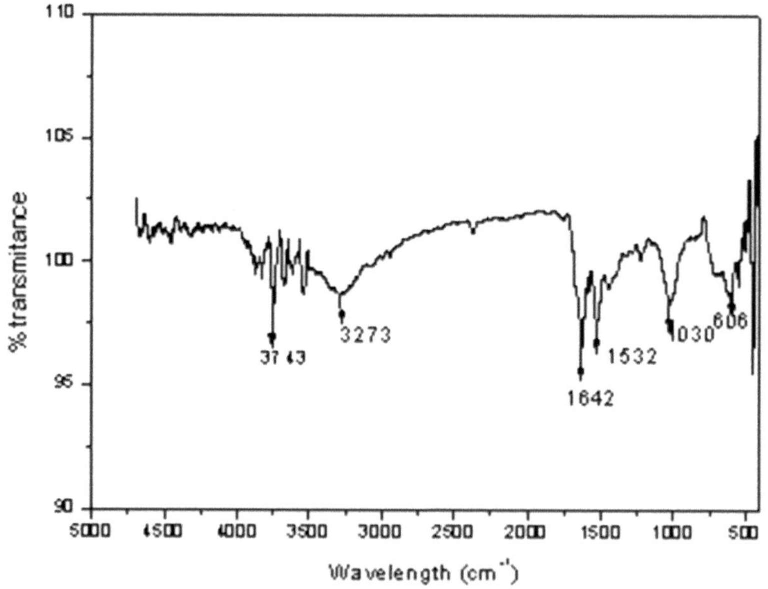

FIGURE 2.19 FTIR spectra of fish scale particulate.

2.4 CONCLUSION

The epoxy reinforced with fish scale is made by hand layup technique. The mechanical characterizations have been carried out and found the tensile strength. The results reveal that the strength is decreasing by increasing the particulate reinforcement. The newly made composites are not transferring the load between matrix and reinforcement due to the unavoidable few bubbles and voids in a specimen with the least bonding between matrix and reinforcement. The maximum tensile strength of 65 Mpa, the compressive strength of 153 Mpa, shore D hardness of 99 SHN, and flexural strength of 355 Mpa were found. This was improved by incorporating E glass fiber single lamina at the bottom and top layer as sandwich polymer composites which have been increased further, adding more layers.

KEYWORDS

- **e-glass fiber**
- **fatigue strength**
- **fish scale**
- **flexural strength**
- **gram per square meter**
- **natural fiber particulates**
- **surface morphologies**
- **woven roving**

REFERENCES

1. Fereydoun, P., Ali, J., Seyedhossein, H., & Azra A., (2014). In vitro and in vivo evaluation of a new nanocomposite, containing high-density polyethylene, tricalcium phosphate, hydroxyapatite, and magnesium oxide nanoparticles. *Materials Science and Engineering C, 40*, 382–388.
2. Se-Kwon, K., & Eresha, M., (2006). Bioactive compounds from marine processing byproducts: A review. *Food Research International, 39*, 383–393.

3. Toshiyuki, I., Hisatoshi, K., Junzo, T., Dominic, W., & Stephen, M., (2003). Microstructure, mechanical, and biomimetic properties of fish scales from Pagrus major. *Journal of Structural Biology, 142*, 327–333.
4. Gil-Duran, S., Arola, D., & Ossa, E. A., (2016). Effect of chemical composition and microstructure on the mechanical behavior of fish scales from *Megalops atlanticus*. *Journal of the Mechanical Behavior of BioMedical Material, 56*, 134–145.
5. Oscar, H. O. N., Carolina, B., & Carlos, E. D., (2017). High-temperature CO_2 capture of hydroxyapatite extracted from tilapia scales. *Journal of the Faculty of Sciences, 22*(3), 215–236.
6. Yi-Cheng, H., Pei-Chi, H., & Huey-Jine, C., (2011). Hydroxyapatite extracted from fish scale effects on MG63 osteoblast-like cells. *Ceramics International, 37*, 1825–1831.
7. Ashley, B., Christine, O., & Mary, C. B., (2013). Mechanics of composite plasmoid fish scale assemblies and their bio-inspired analogues. *Journal of the Mechanical Behavior of Biomedical Material, 19*, 75–86.
8. Nawshad, M., Girma, G., Abdur, R., Pervaiz, A., Farasat, I., Faiza, S., Amir, S. K., et al., (2017). Investigation of ionic liquids as a pre-treatment solvent for extraction of collagen biopolymer from waste fish scales using COSMO-RS and experiment. *Journal of Molecular Liquids, 232*, 258–264.
9. Alberto, C., Egidio, R., & Enrico, P., (2000). Experimental characterization and numerical simulations of a syntactic-foam/glass-fiber composite sandwich. *Composites Science and Technology, 60*, 2169–2180.
10. Mostafa, A., Shankar, K., & Morozov, E. V., (2014). Behavior of PU-foam/glass-fiber composite sandwich panels under flexural static load. *Materials and Structures, 253(3)*, 11527.

CHAPTER 3

A Review on Past and Future Aspects of Silica in Drug Delivery and Sensing Applications

HARPREET KAUR and GAGANDEEP KAUR

Department of Chemistry, Punjabi University, Patiala–147002, Punjab, India, E-mail: preetjudge@yahoo.co.in (H. Kaur)

ABSTRACT

The present chapter highlights the significant role played by silica particles in their nano as well as micro dimensions in advanced research fields. It is observed that alternation in surface properties of silica or combination with suitable organic/inorganic groups enhance its drug holding capacity, biocompatibility, multifunctionality, and stability under a large pH range. For a better understanding of this chapter, chronological developments in the synthesis of silica from tetraethyl orthosilicate (TEOS) and rice plant waste have been discussed first. Secondly, the applicative area of silica covering electrochemical sensors, drug delivery carriers, and optical sensors, along with their future aspects, has also been discoursed.

3.1 INTRODUCTION

Silica in its nano form possesses great potential in sensing, biomedical, and theranostics research due to its high stability, low toxicity, and its ability to get functionalized with polymers, amino acids, and proteins of biological importance [1]. Nano silica/silica nanoparticles are divided

Advances in Diverse Applications of Polymer Composites: Synthesis, Application, and Characterization. Suji Mary Zachariah, Yang Weimin, Maciej Jaroszewski, & Sabu Thomas (Eds.)

into P-type or S-type according to their structural properties. P-type nanoparticles are porous, and show higher ultraviolet reflectivity and large surface area as compared to that spherical S-type silica nanoparticles. Silica is biocompatible and relevantly easily available in the environment as compared to oxides of other metals and non-metals. Silica may also be found in combination with alkali and alkaline earth metal ions, which are known as silicates or can be solely existed in polymerized form. Although silica exists in numerous polymorphs yet, only three principal forms of silica are of great importance and are potentially studied. Out of these three polymorphs, quartz is the amorphous form of silica, whereas cristobalite and tridymite are two temperature-dependent crystalline forms of silica [2]. At pH 7 and 25°C, the solubility of silica in water is 120 mg/l, which increases with an increase in temperature. Silica exhibits versatile properties due to its unique individuality like; biodegradability, biocompatibility, high pore volume, large surface area, and free dispersion throughout the body. Most importantly, it allows homogenous distribution of guest molecules in available porous voids [3]. The exceptional specialty of SiO_2 lies in the fact that its physical and chemical properties can be altered as per requirements by precisely optimizing experimental conditions as well as by functionalization of silanol groups present on its upper surface. Because of strong Si-O bonding, silica can tolerate high mechanical stress as compared to liposomes, niosomes, and dendrimers [4]. Silica is mostly utilized in integrated circuits, silicon-based device fabrication technology, food additives, sensors, adsorption, fillers, and extenders in concrete, glass manufacturing, and drug delivery [5–16]. Mesoporous silica nanoparticles (MSNs), due to low toxicity and high drug loading capacity, are highly recommended to control and target drug delivery systems in human as well as animal bodies [17]. MSNs have well-defined and tunable structural properties, which can be altered via functionalization of the silanol group for better drug loading and sustainable release as per medical requirements. MSNs are also popularly recommended for gene transport, gene expression, biosignal probing, imaging agent, detecting agent, drug delivery vehicles, etc. [18]. In addition to the insulating properties of silica, it shows ubiquitous applications as electrochemical as well as biosensors. Recently, silica-based materials have been reviewed as electrochemical sensor, biosensor [19], genosensors or immunosensors [20], potentiometric sensors [21], gas detectors [22], electrochemiluminescence sensors [23], etc. For electrochemical sensing, nano-silica is dispersed with conducting polymer to form nanocomposites or deposited as nano labels/films on electrode surfaces [24]. Unique features of silica can be scaffolded to protect biological objects on electrodes, and new avenues in the field of biosensors can be developed. Mesoporous silica-based electrochemical sensors and biosensors with ordered morphology show improved performance as compared to that non-ordered analogous materials. Silica nanoparticles are mostly used in the fabrication of pH-based sensors.

3.2 SYNTHESIS OF SILICA

Silica has been reported to be synthesized from tetraethyl orthosilicate/ tetraethoxysilane (TEOS) using the sol-gel technique in nano as well as micro dimensions. Sol-gel technique provides accessible tuning in the composition, morphology, and topology of synthesized particles. It has been reported that nano-silica can also be extracted from agrowastes by adopting an acid-alkaline treatment/leaching process. In this chapter, two protocols of synthesis of silica, i.e., one using TEOS and the other from agro-wastes (rice straw (RS) and rice husk), along with chronological developments in synthesis' procedure and the resulting desirable alternations in properties as well as in applicative domain have been discussed. The principal focus is highlighted on the utility of silica in the field of electrochemical sensing, drug delivery vehicles, diagnosis, therapeutics, and optical biomedical imaging.

3.2.1 SYNTHESIS OF SILICA USING TETRAETHYL ORTHOSILICATE (TEOS)

Tetraethyl orthosilicate (TEOS) with the chemical formula $Si(OC_2H_5)_4$ is the ethyl ester of orthosilicic acid with a sharp alcohol-like smell. It plays the role of cross-linking agent and is popularly recommended as a precursor to silicon dioxide (SiO_2) in the laboratory and semiconductor industry.

Sol-gel technique provides controlling hand on size distribution range even at low temperature. In the sol-gel process, the metal or non-metal salt in a polar solvent (alcohols) undergoes hydrolysis and poly-condensation reactions to create a gel-type colloidal suspension having size morphologies ranging from discrete particles to polymeric networks. A gel-like suspension converts into powder form upon heating, which contains an oxide of a metal or non-metal [25]. The following section represents the yearly developments in the methodology of silica production using TEOS as a precursor to silica, along with the adopted spectroscopic techniques for characterization purposes.

Werner Stöber had the honor to first synthesize spherical-shaped monodispersed silica particles in 1968 in the colloidal range (50–2,000 nm) using TEOS as a precursor for silica generation followed by condensation of silicic acid in alcoholic solutions (methanol, ethanol, n-propanol, and n-butanol) in the presence of ammonia as a morphological catalyst. The surface morphology and particle size range of synthesized silica particles have been characterized using scanning electron microscopy (SEM) and the dynamic light scattering (DLS) method. This method was based on the sol-gel technique of synthesis [26].

In 2005, Rao et al. synthesized silica nanoparticles by sol-gel method followed by ultrasonication. It has been observed that particle size decreases with an increase in the concentration of reagents. The researchers characterized these particles using various analytical techniques. The electronic absorption behavior of silica nanoparticles has also been studied using UV-visible spectroscopy [27].

Razo et al. in 2008, prepared spherical-shaped monodispersed silica of 400 nm size for the fabrication of opal photonic crystals in one-pot synthesis by carefully maintaining the concentration of TEOS, NH_3, and H_2O [28].

During 2009, Thomassen et al. synthesized silica sols (2–335 nm) for testing the *in vitro* cytotoxicity. The particle size, surface morphology, porosity, and surface area have been characterized using DLS, SEM, and nitrogen adsorption. The relation between particle size of silica particles and cytotoxicity has been drawn using human endothelial and mouse monocyte-macrophage cells. The results highlighted the strong connection between cytotoxicity and the size of silica particles with cell type [29].

Murray et al. [30] compared the various routes of synthesis of silica particles and suggested a novel and direct approach to prepare hydrophobic organosilica nanoparticles using octadecyltrimethoxysilane (ODTMS) as single silica precursor [30]. Wang et al. has been successfully synthesized silica particles in the size range of 20–1,000 nm by following Stöber's method but using high concentration of TEOS. After carefully investigating the influence of TEOS, ammonia, and water on particle size distribution, a modified monomer addition model has been combined with aggregation model to study the possible mechanism of formation of silica particles in 2010 [31].

In 2011, Canton et al. synthesized dye-doped nano-silica for photocatalytic stability and ultra-sensitivity, which can be further utilized in biosensing and bioimaging. Two methodologies have been followed, i.e., Stöber's method and the microemulsion method. Out of these two, Stober's protocol gave better results when the reaction was carried out with 3-Aminopropyl-triethoxysilane (APTES) linked to Alexa Flour 555 dye. Particles in the size range of 10–50 nm have shown superior fluorescence lifetime and biomedical applications as compared to that of large size particles [32].

Ganesh et al. [33] followed sol-gel approach for the synthesis of MSNs with high surface area and pore volume by using Triton X-100 as main template and Tween 60 as co-template. These MSNs have been loaded with water-insoluble drug, i.e., ibuprofen. Drug loading capacity and sustainable release capacity has been examined for ibuprofen encapsulated on MSNs [33]. Micelles entrapment approach has been used by Zainal et al. [34] for the synthesis of spherical nano-silica. Effect of various parameters, such as pH, stirring time, and reagents' concentration has been studied. It has been found that by adjusting reaction temperature and proportion of reactants, silica nanoparticles can be synthesized in size range of 28.91 nm to 113.22 nm [34].

Magnetic mesoporous silica composites (120–380 nm) have been synthesized in 2014 by Wang et al. by incorporation of magnetic Fe_3O_4 using modified Stöber's method and cetyltrimethylammonium bromide (CTAB) as surfactant. These particles showed good biomedical adaptation performance and thus have been utilized for drug delivery platforms and photodynamic therapy [35]. Singh et al. [36] synthesized silica/graphene oxide (GO) nanocomposites by modified Stöber's method and Hummer's method, which have been categorized using FTIR, XRD, TEM (transmission electron microscopy), and SEM. GO plays the role of reinforcement material and improves the mechanical properties, electrical conductivity, and adsorption properties of silica [36]. In 2015, Delyan and his co-workers described the method to control the size homogeneity in silica nanoparticles by two-phase arginine catalyzed method through the organic solvent phase [37].

Zulfiqar et al. [38] reported a method to develop silica particles by acid and thermal treatment of Bentonite clay. Dimensions of silica particles have been varied from nano to micro by changing the contents of silica-rich clay, ethanol, and nitric acid. It has been observed that the concentration of silica-rich clay and nitric acid directly affects particle size whereas an increase in ethanol quantity produces bimodal particles of nano as well as micro dimensions [38].

Chemical methods of synthesis of silica particles either in micro or nano dimensions using TEOS require harsh conditions and economically as well as environmentally unfavorable chemicals. These methods are time depleting also.

3.2.2 BIOGENIC EXTRACTION OF SILICA FROM RICE PLANT WASTE

Because of the demerits of the above-mentioned protocols of synthesis of silica, the interest of researchers has uplifted towards the natural sources of silica, which will be environment friendly, economically feasible, and will utilize easily available chemicals. To attain these objectives, biogenic methods of synthesis of silica have been proposed in which silica is basically extracted from easily available natural resources. Rice is the second most utilized crop all over the globe. During the cultivation of rice, paddy, RS, and rice husk (RH) are the three major obtained ingredients. Paddy is utilized for eating purposes, but RH and RS are the waste products having no utility. Here, the extraction of silica from RS and RH will be discussed for their vide applications in the various sectors.

Three-step oxidation treatment of RS using $NaClO_2$, toluene/ethanol, and KOH (potassium hydroxide) has been employed to extract silica particles from it. Porous spherical-shaped silica has been precipitated by using acidified poly (ethylene oxide) and calcinated at 500°C in pure form with particle size ranging from 100–120 nm [39].

Silica disks in pure amorphous form have been obtained by refluxing rice straw ash (RSA) which has been obtained by heating RS at 575°C. Refluxing is carried out in with 0.5 M NaOH for 5 hours at 100°C. The sodium silicate slurry has been neutralized with sulfuric acid to gel mesoporous silica nanodisks with average pore size of 5.8 nm (2–22 nm pore size distribution), high specific surface (BET (Brunauer-Emmett-Teller) surface of 509.5 m^2/g and BJH cumulative surface of 637.0 m^2/g) and pore volume (0.925 cm^3/g) [40].

Silica particles have been extracted along with lignin by adopting two-step pretreatment process from paddy straw agro-waste. It has been given acid washing with H_2SO_4 followed by delignification with mixture of H_2O_2 and NaOH at 100°C for 3 hours. Accumulated agro-residue waste has been neutralized by acidifying with H_2SO_4 at 37°C. Silica gel was obtained by adjusting pH to 6.5 and dried overnight in an oven to get silica particles [41].

Under properly maintained anaerobic conditions, fragments of RS of length 10–50 mm have been refluxed with HCl at 90C for 1 hour. The treated RS fragments have been burnt at 300°C to remove volatile substances and to obtain RSA, which has been further calcinated at 600–700°C till a constant mass is obtained. This sample has been identified by FTIR and XRD and is found to have amorphous silica particles with impurities due to originally present impure organic species in the corresponding plant variety of RS [42].

Semi-crystalline silica nanoparticles have been successfully synthesized by boiling RH for 2 hours in 10% HCl. It has been washed and pyrolyzed at 700°C for 2 hours to obtain rice husk ash (RHA). RHA has been ultrasonicated and stirred in 0.20 M KNO_3. Dried in oven and filtrate has been calcinated at 800°C for 4 hours, which provided crystalline silica nanoparticles [43].

In another report, RH has been grounded into a fine powder, which has been mixed with ionic liquid for extraction of lignocelluloses. The dried lignocelluloses have been pyrolyzed for 2 hours at 700°C in order to produce silica nanoparticles [44].

RHA has been obtained by 4-hour thermal treatment of Vietnamese rice husk at 600°C. Further, the sol-gel technique has been used for extraction of silica from RH in which NaOH, H_2SO_4 (30%), HCl (10%), and water/butanol solvent systems are employed. For optimizing the homogeneity of silica nanoparticles, CTAB is used as a surfactant. BET studies revealed that amorphous nano-silica powder with a high surface area of 340 m^2/gm had been successfully obtained [45].

The major drawback of the above-mentioned methods is the combustion of RS and RH into their ashy material. Because it requires a very high temperature and toxicant gases are emerged in the atmosphere due to combustion. These are ultimately responsible for the generation of air and water pollution. Moreover, very harsh chemicals have been used. Thus, there is a need to find an alternative approach to the synthesis of silica, which will be in favor of the global beneficiary.

3.3 APPLICATIONS OF SILICA

Ordered mesoporous nanoparticles have been discovered two decades ago. Nowadays, these are gaining a lot of attention due to their unique properties of large surface area, high pore volume, tunable size, wide range of pore size, easily approachable active sites, mesoporous channels, and hosting capabilities, which make these widely used in development of electrochemical sensor, biosensor [19], genosensors or immunosensors [20], potentiometric sensors [21], gas detectors [22], electrochemiluminescence sensors [23]. The three major applications of silica are electrochemical sensors, drug delivery carriers, and optical sensors.

3.3.1 ELECTROCHEMICAL SENSORS

Electrochemical sensors are widely used in integrated electrical devices. These sensors record any change in properties, like potential, current density, charge on surface, temperature, etc., and after amplification develop them as signal [46]. Silica-based nanocomposites hold the ability to act as sensors because of the silanol groups present on its surface. These outer silanol groups have the affinity to get functionalized with foreign materials (conducting polymers or metal/metal oxide/mixed metal oxide nanoparticles), which altogether provide a platform for sensing activity. Change in current, potential or resistance has been observed due to the interactions between silica-based nanocomposites and targeting entity, and it is appeared as signal. Series of signals are plotted in graphical form to precisely discuss the obtained results [47].

Core-shell nanobeads and nanostructured MSNs have gained attention in electrochemical sensing. In silica-based core-shell nanoparticles, silica can duly act as:

- In the first category, silica can act as core, which is supporting a reactive shell onto its surface and it is generally symbolized as (SiO_2@shell). These shells are mostly grafted on the surface of silica via in situ reactions by functionalization through attached silanol groups. For instance, conducting polymer, such as polyaniline (PANI) has been polymerized on nano-silica core (SiO_2/PANI) for sensing of ascorbic acid [48]. Gold nanoparticles (AuNPs) act as shell on silica core (SiO_2/AuNPs) for detection of acetylsalicylic acid [49]. Silica nanoparticles' core when covered with molecularly imprinted polymer (MIP) shells, act as an electrochemical sensor for 2,6-aminopyridine in hair dyes after deposition onto glassy carbon electrode (GCE) [50].

- In the second category, silica can act as shell around other metal or metal oxide core, which is noted as (core@SiO_2). Here, silica acts as a protector for metal or metal oxide, provides surface for further functionalization with organic groups, etc. Magnetically active nanoparticles in silica shells with immobilized biomolecules provide a hybrid platform for reagentless enzymatic biosensors [51].

SBA-15 and SBA-16 have been successfully fabricated by using TEOS as silica precursor and Pluronic P123 and F123 surfactants as directing agents [52]. These have been combined with modified carbon paste electrodes (MCPE) and used to study the electrochemical behavior of ascorbic acid and uric acid via cyclic voltammetry [53].

Mesoporous silica thin film modified GCE has been reported to detect cationic paraquat in an aqueous medium via cyclic voltammetry. This is because of the well-organized and oriented nano-dimensional pores and mesochannels present in mesoporous silica, which vary with change in pH, ionic strength, nature of supporting electrolyte anion, etc. With this method limit of detection (LOD) in real samples was found to be spiked, i.e., 12 nM, which is found to be three times better than already reported LOD (40–400 nM) for drinking water [54].

3.3.2 DRUG DELIVERY

In the past times, drug encapsulation and drug release have been based on conventional methods, which seriously suffer from cons, like less solubility, dose toxicity, and short half-life. To overcome these disadvantages, a nanoparticle-based drug delivery system has been invented. In this method, the drug has been encapsulated around nanoparticles for better target action, sustained action, drug protection, and higher bioavailability [55]. Silica nanoparticles in mesoporous dimensions (5–50 nm) have been considered the most suitable for drug delivery based on nanoparticles. MSNs have versatile drug delivery systems because of their better guest-holding properties, such as peptides, proteins, anticancer agents, and genetic materials [56]. Thoroughly loading and timely release of drug can only be achieved by mesoporous silica due to its structure homogeneity and controllable pore size, which is required by biocompatible guest molecules. Solubility of silica is around 2–3 mM at intestinal pH [57]. Basically, silica plays a pivotal role as a carrier of various drugs and their substantial release in the body without imparting any harmful effect on other organs and disturbing metabolic processes as it excretes out of the body as such. Mesoporous silica has acted as an oral solid dispersion carrier for hydrophobic cilostazol [58]. It has also enhanced the dissolution of indomethacin nanoparticles and thereby reduced gastric irritancy [59]. Phosphonate MSNs (45 nm) have been reported for the therapeutic treatment of various kinds of cancers due to superior *ex vitro*, *in vitro*, and *in vivo* drug elution and targeted delivery performance. This is because of efficient passive targeting, intratumoral diffusion, intracellular retention, and fast controlled release of such small phosphonate nanoparticles [60]. It has been investigated that the particle size of MSNs and their elution performances are the key points for the development of tailored-made nano-vectors for rising applications in the nanomedicine field.

Design of MSNs and loading of drug on the surface of MSNs should be such that it can respond to changes in specific stimuli, like pH variations, redox potential, glutathione concentration, nature of enzyme, and presence of several other molecules [51]. The smart designing of MSNs includes capping agent/linker (hinders premature cargo departure) and blocking cap (able to degrade under stimulus action and permits uncapping of drug for sustainable release). Stimuli-responsive strategy for smart drug delivery MSNs is summarized in Table 3.1.

3.3.3 OPTICAL SENSORS

Light is the most intuitive tool for creatures to recognize the world. Luminescent materials are used for lighting, displaying, and changes in their fluorescent emission intensity can directly provide a pathway to derive some important information. The basic art of luminescent sensors has been demonstrated by Michael Schäfrling [84]. Silica, being transparent to light, acts as an ideal matrix for versatile phosphors. Luminescent materials suffering from demerits, such as low hydrophobicity, less absorbance, poor biocompatibility, toxic behavior, etc., can be modified by incorporating with silica matrix because of its greater thermodynamic

TABLE 3.1 Stimuli-Responsive Strategy for Smart Drug Delivery by MSNs

Stimulus	Capping Agent or Linker	Blocking Agent	References
pH	Acetal linker	AuNPs	[61]
pH	Bromate ester	Fe_3O_4NPs	[62]
pH	Ferrocenyl moieties	β-CD modified CeO_2NPs	[63]
pH	PAH-PSS-PEM	PAH-PSS-PEM	[64]
pH	Aromatic amines	CDs	[65]
pH	Benzoic-imine bonds	Polyseudorotaxanes	[66]
pH	Soluble CaP	CaP coating	[67]
pH	Self-immolation polymer	Self immolation polymer	[68]
pH	Gelatin	Gelatin coating	[69]
pH	3,9-Bis(3-aminopropyl)-2,4,8,10-tetraoxaspiro [5.5] undecane (ATU)	Poly(acrylic acid)	[70]
Redox potential	–S–S–	ssDNA	[71]
Redox potential	–S–S–	PEG	[72]
Redox potential	–S–S–	CdS NPs	[73]
Redox potential	–S–S–	PPI dendrimer	[74]
Enzyme	MMP-degradable gelatin	Gelatin coating	[75]
Enzyme	β-galactosidase cleavable oligosaccharide	β-galacto-oligosaccharide	[76]
Enzyme	MMP9-sensitive peptide sequence	Avidyne	[77]
Enzyme	protease-sensitive peptide sequences (CGPQGIWGQGCR)	PNIPAm-PEGDA shell	[78]
Enzyme	α-amylase and lipase cleavable stalks	CDs	[79]
Enzyme	Phosphate-phosphate APasa-hydrolyzable bonds	ATP (adenosine triphosphate)	[80]
Small molecules	Ionizable benzimidazole group	CD modified glucose oxidase	[81]
Small molecules	pAb	pAb	[82]
Small molecules	ATP aptamer	ATP aptamer	[83]

stability, chemical versatility, and tunable surface properties via functionalization to adapt diverse requirements [85]. Luminescent metals, organic dyes, and quantum dots doped phosphors are the three main phosphors, which are highly proficient to embed into silica matrix to act as luminescent silica nanoparticles (LSNs). Phosphors have been doped into silica matrix to improve their properties. LSNs possess many applications in the field of biolabeling, bioimaging, diagnosis, solar cells, and photocatalytic activity [23].

An18/SiO_2NPs has been prepared by encapsulation of An18 (derived from 9,10-distyrylanthracene with an alkoxyl end group) on silica nanoparticles via the one-pot modified Stöber method. These modified LSNs showed enhanced fluorescence intensity, non-toxicity toward living cells, good water solubility, and improved potential for biomedical applications [86]. Shi et al. studied europium-based MSNs and used their fluorescent property for biolabeling and tissue repairing. It has been found that by promoting proper response of macrophages and expression of relevant genes, bone defects can be replaced, and healing of skin can also be accelerated with Eu-MSNs [87]. Silica-based phosphors, their composition, and the protocol of synthesis are given in Table 3.2.

3.4 CONCLUSION

Silica is a highly important inorganic material whose various applications can be studied by tuning its properties because of the presence of silanol groups on its surface. Its applications in concrete fillers, adsorption, abrasive/anticaking agents, and drug delivery carriers/theranostics have been explored very well. But its potential implementation as an electrochemical sensor and optical sensor is still under the canvas of investigation. Thus, there is a need to develop silica-based nanocomposites which can act as proficient electrochemical as well as biochemical sensors. All together this, many new luminescent materials with excellent performance have been developed, but the great potential in the field of LNSs still needs to be investigated. Porous hollow silica nanoparticles based on controlled release of pesticides on targeted crops in the presence of template or co-template can be a better idea for innovative research. Intensive crop cultivation decreases the silicon content in the soil, which ultimately declines the crop's yield and fertility of the soil. Treatment of soil with

TABLE 3.2 Luminescent Silica Nanoparticles-based Phosphors, Their Composition, and Protocol of Synthesis

Luminescent Silica Nanoparticles (LSNs)	Composition	Protocol of Synthesis	References
Organic moiety doped LSNs	Y_2O_3:Eu_3+@SiO_2 with FITC	Stöber method	[88]
	Amino cyanine dye-silica hybrid nanoparticles	Reverse microemulsion	[89]
	AIE-F127-SiO_2	Sol-gel method	[90]
	Rhodamine-conjugated silica	Direct micelles assistant method	[91]
Metal doped LSNs	Eu@Si-OH and Eu@Si-NH_2	Reverse microemulsion	[92]
	$NaGdF_4$:Yb,Er@SiO_2@Eu $(TTA)_3$Phen	Reverse microemulsion	[93]
	Silica@EuCP	Solvothermal method	[94]
	Ru $(bpy)_3$ doped silica	Reverse microemulsion	[95]
	Ru $(bpy)_3$@SiO_2	Stöber method	[96]
	Er,Yb:$GdVO_4$@SiO_2	Stöber method	[97]
Quantum dots-doped LSNs	CdSe/CdS/ZnS@SiO_2	Stöber method	[98]
	Quantum dot/SiO_2/Au	Reverse microemulsion	[99]
	Silica encapsulated polymer dots	Stöber method	[100]

silica can complete its silicon deficiency, and the uptake of nutrients by plants will eventually be increased. Therefore, silica-based materials need to be investigated more and more by researchers to resolve such kinds of problems for the betterment of the future.

3.5 FUTURE ASPECTS

From the above discussion, it is clear that silica has been extracted either from TEOS or by using highly concentrated chemicals and by proceeding with the ash of plant waste. These methods are opposite to the norms of a healthy economic state. Thus, a new environment-friendly methodology can be proposed, which will proceed without the combustion of agro-waste into ash material and with the use of a very small quantity of chemicals. Instead of using spectrophotometric and cyclo-voltammetric sensors for the detection of metal ions, potentiometric sensors based on silica membranes can be developed for easy and economical use. In the future, research can be explored the use of silica-based materials as carriers for the controlled release of pesticides and skin ointments. This can also be extended to enhance the yield of healthy crops by increasing the silica uptake content of the soil, i.e., to increase the fertility of the soil. Although these innovative ideas are not explored, the research in this field can bring a revolution in pesticide delivery as well as the cosmetic industry.

KEYWORDS

- **drug delivery**
- **dynamic light scattering**
- **electrochemical sensor**
- **luminescent silica nanoparticles**
- **mesoporous silica nanoparticles**
- **optical sensor**
- **silica**

REFERENCES

1. Mebert, A. M., Baglole, C. J., Desimone, M. F., & Maysinger, D., (2017). Nanoengineered silica: Properties, applications and toxicity. *Food Chem. Toxicol., 109*, 753–770.
2. Nabila, M., Mahmoudb, K. R., El-Shaerb, A., & Nayberb, H. A., (2018). Preparation of crystalline silica (quartz, cristobalite and tridymite) and amorphous silica powder (one step). *J. Phys. Chem. Solids, 121*, 22–26.
3. Guo, C., Wang, J., Yang, M., Li, Y., Cui, S., Zhou, X., Li, Y., & Sun, Z., (2017). Amorphous silica nanoparticles induce malignant transformation and tumorigenesis of human lung epithelial cells via P53 signaling. *Nanotoxicology, 11*, 1176–1194.
4. Singh, A. P., Biswas, A., Shukla, A., & Maiti, P., (2019). Targeted therapy in chronic diseases using nanomaterial-based drug delivery vehicles. *Signal Transduct. Tar., 4*. doi: 10.1038/s41392-019-0068-3.
5. Villota, R., Hawkes, J. G., & Cochrane, H., (2009). Food applications and the toxicological and nutritional implications of amorphous silicon dioxide. *Crit. Rev. Food Sci. Nutr., 23*, 289–321.
6. Hadipramana, J., Riza, F. V., Rahman, I. A., Loon, L. Y., Adnan, S. H., & Zaidi, A. M. A., (2016). Pozzolanic characterization of waste rice husk ash (RHA) from Muar, Malaysia. *IOP Conf. Ser.: Mater. Sci. Eng., 160*. doi: 10.1088/1757-899X/160/1/012066.
7. Riberio, T., Raja, S., Rodrigues, A. S., Fernandes, F., Baleizão, C., & Farinha, J. P. S., (2014). NIR and visible perylene diimide-silica nanoparticles for laser scanning bioimaging. *Dyes and Pigments, 110*, 227–234.
8. Wang, Y., Zhao, Q., Han, N., Bai, L., Li, J., Liu, J., Che, E., et al., (2014). Mesoporous silica nanoparticles in drug delivery and biomedical applications. *Nanomedicine: Nanotech. Bio. and Med.* doi: 10.1016/j.nano.2014.09.014.
9. Geng, J., Li, M., Wu, L., Chen, C., & Qu, X., (2012). Mesoporous silica nanoparticle-based H_2O_2 responsive controlled-release system used for Alzheimer's disease treatment. *Adv. Healthcare Mater., 1*, 332–336.
10. Jaroenworaluck, A., Sunsaneeyametha, W., Kosachan, N., & Stevens, R., (2006). Characteristics of silica-coated TiO_2 and its UV absorption for sunscreen cosmetic applications. *Surf. Interface Anal., 38*, 473–477.
11. Bailly, M., Kontopoulou, M., & Mabrouk, K. E., (2010). Effect of polymer/filler interactions on the structure and rheological properties of ethylene-octene copolymer/nanosilica composites. *Polymer, 51*, 5506–5515.
12. Rahman, I. A., & Padavettan, V., (2012). Synthesis of silica nanoparticles by sol-gel: Size-dependent properties, surface modification and applications in silica-polymer nanocomposites: A review. *J. Nanomater., 2012*, 1–15.
13. Sun, L., & Gong, K., (2001). Silicon-based materials from rice husks and their applications. *Ind. Eng. Chem. Res., 40*, 5861–5877.
14. Dukarska, D., & Czamecki, R. E., (2016). Fumed silica as a filler for MUPF resin in the process of manufacturing water-resistant plywood. *J. Wood Prod., 74*, 5–14.
15. Samiey, B., & Toosi, A. R., (2010). Adsorption of malachite green on silica gel: Effects of NaCl, pH and 2-propanol. *J. Hazard. Mater., 184*, 739–745.
16. Zhu, W., Wang, J., Wu, D., Li, X., Luo, Y., Han, C., & Ma, W., (2017). Investigating the heavy metal adsorption of mesoporous silica materials prepared by microwave synthesis. *Nanoscale. Res. Lett., 12*, 1–9.

17. Douroumis, D., Onyesom, I., Maniruzzaman, M., & Mitchell, J., (2012). Mesoporous silica nanoparticles in nanotechnology. *Crit. Rev. Biotechnol.,* 1–17. doi: 10.3109/07388551.2012.685860.
18. Zhou, Y., Quan, G., Wu, Q., Zhang, X., Niu, B., Wu, B., Huang, Y., et al., (2018). Mesoporous silica nanoparticles for drug and gene delivery. *Acta Pharm. Sin. B., 8,* 165–177.
19. Walcarius, A., (2018). Silica-based electrochemical sensors and biosensors: Recent trends. *Curr. Opin. Electrochem., 10,* 88–97.
20. Kimmel, D. W., LeBlanc, G., Meschievitz, M. E., & Cliffel, D. E., (2012). Electrochemical sensors and biosensors. *Anal. Chem., 84,* 685–707.
21. Safronovaa, E., Parshinab, A., Kolganovab, T., Bobreshovab, O., Pourcellyc, G., & Yaroslavtseva, A., (2018). Potentiometric sensors arrays based on perfluorinated membranes and silica nanoparticles with surface modified by proton-acceptor groups, for the determination of aspartic and glutamic amino acids anions and potassium cations. *J. Electroanal. Chem., 816,* 21–29.
22. Sun, L., Semenova, Y., Wu, Q., Liu, D., Yuan, J., Sang, X., Yan, B., et al., (2017). High sensitivity ammonia gas sensor based on a silica gel coated microfiber coupler. *J. Lightwave Technol., 35,* 2864–2870.
23. Li, L., Wang, W., Tang, J., Wang, Y., Liu, J., Huang, L., Wang, Y., et al., (2019). Classification, synthesis and application of luminescent silica nanoparticles: A Review. *Nanoscale Res. Lett., 14,* 1–23.
24. Ibanez, J. G., Rincón, M. E., Gutierrez-Granados, S., Chahma, M., Jaramillo-Quintero, O. A., & Frontana-Uribe, B. A., (2018). Conducting polymers in the fields of energy, environmental remediation and chemical–chiral sensors. *Chem. Rev., 118,* 4731–4816.
25. Le, V. H., Thuc, C. N. H., & Thuc, H. H., (2013). Synthesis of silica nanoparticles from Vietnamese rice husk by sol-gel method. *Nanoscale Res. Lett., 8,* 1–10.
26. Stöber, W., Fink, A., & Bohn, E., (1968). Controlled growth of monodisperse silica spheres in the micron size range. *J. Colloid Interface Sci., 26,* 62–69.
27. Rao, K. S., El-Hami, K., Kodaki, T., Matsushige, K., & Makino, K., (2005). A novel method for synthesis of silica nanoparticles. *J. Colloid Interface Sci., 289,* 125–131.
28. Razo, D. A. S., Pallavidino, L., Garrone, E., Geabaldo, F., Descrovi, E., Chiodoni, A., & Giorgis, F., (2008). A version of Stöber synthesis enabling the facile prediction of silica nanospheres size for the fabrication of opal photonic crystals. *J Nanopart Res., 10,* 1225–1229.
29. Thomassen, L. C. J., Aerts, A., Rabolli, V., Lison, D., Gonzalez, L., Kirsch-Volders, M., Napierska, D., et al., (2010). Synthesis and characterization of stable monodisperse silica nanoparticle sols for in vitro cytotoxicity testing. *Langmuir, 26,* 328–335.
30. Murray, E., Born, P., Weber, A., & Kraus, T., (2010). Synthesis of monodisperse silica nanoparticles dispersible in non-polar solvents. *Adv. Eng. Mater., 12,* 374–378.
31. Wang, X. D., Shen, Z. X., Sang, T., Chenga, X. B., Li, M. F., Chen, L. Y., & Wang, Z. S., (2010). Preparation of spherical silica particles by Stöber process with high concentration of tetra-ethyl-orthosilicate. *J. Colloid Interface Sci., 341,* 23–29.
32. Canton, G., Riccò, R., Marinello, F., Carmignato, S., & Enrichi, F., (2011). Modified Stöber synthesis of highly luminescent dye-doped silica nanoparticles. *J. Nanopart Res., 13,* 4349–4356.

33. Ganesh, M., & Lee, S. G., (2013). Synthesis, characterization and drug release capability of new cost effective mesoporous silica nanoparticle for ibuprofen drug delivery. *Int. J. Control Autom., 6*, 207–216.
34. Zainala, N. A., Shukor, S. R. A., Wabb, H. A. A., & Razak, K. A., (2013). Study on the effect of synthesis parameters of silica nanoparticles entrapped with rifampicin. *AIDIC Conference Series, 11*, 431–440.
35. Wang, F., Li, Z., Liu, D., Wang, G., & Liú, D., (2014). Synthesis of magnetic mesoporous silica composites via a modified Stöber approach. *J. Porous Mater.*, 513–519.
36. Singh, V. K., Cura, M. E., Liu, X., Johansson, L. S., Ge, Y., & Hannula, S. P., (2014). Tuning the mechanical and adsorption properties of silica with graphene oxide. *Chem. Plus Chem., 79*, 1512–1522.
37. Hristove, D. R., Mahon, E., & Dawson, K. A., (2015). Controlling aqueous silica nanoparticle synthesis in the 10–100 nm range. *Chem. Commun., 51*, 17420–17423.
38. Zulfiqar, U., Subhani, T., & Husain, S. W., (2016). Synthesis and characterization of silica nanoparticles from clay. *J. Asian Ceram. Soc., 4*, 91–96.
39. Hu, S., & Hsieh, Y. L., (2014). Preparation of activated carbon and silica particles from rice straw. *ACS Sustain. Chem. Eng., 2*, 726–734.
40. Lu, P., & Hsieh, Y. L., (2012). Highly pure amorphous silica nano-disks from rice straw. *Powder Tech., 225*, 149–155.
41. Kauldhar, B. S., & Yadav, S. K., (2018). Turning waste to wealth: A direct process for recovery of nano-silica and lignin from paddy straw agro-waste. *J. Clean. Prod.* doi: 10.1016/j.jclepro.2018.05.136.
42. Zemnukhova, L. A., Panasenko, A. E., Artem'yanov, A. P., & Tsoy, E. A., (2015). Dependence of porosity of amorphous silicon dioxide prepared from rice straw on plant variety. *BioResources, 10*, 3713–3723.
43. Weining, W., Jarett, C. M., Xiotian, F., Aijie, H., Zhiping, L., & Luyi, S., (2012). Silica nanoparticles and frameworks from rice husk biomass. *Appl. Mater. Interfaces, 4*, 977–981.
44. Haoran, C., Weixing, W., Jarett, C. M., Adam, J. O., Paige, A. D., Jeffery, F. X., DeBorn, K. M., et al., (2013). Extraction of lignocellulose and synthesis of porous silica nanoparticles from rice husk: A comprehensive utilization of rice husk biomass. *ACS Sustain. Chem. Eng.*, 254–259.
45. Vallet-Regí, M., Colilla, M., Izquierdo-Barba, I., & Manzano, M., (2018). Mesoporous silica nanoparticles for drug delivery: Current insights. *Molecules, 23*. doi: 10.3390/molecules23010047.
46. Rhazi, M. E., Majid, S., Elbasri, M., Salih, F. E., Oularbi, L., & Lafdi, K., (2018). Recent progress in nanocomposites based on conducting polymer: Application as electrochemical sensors. *Int. Nano Lett., 8*, 79–99.
47. Salis, A., Setzu, S., Monduzzi, M., & Mula, G., (2011). Porous silicon-based electrochemical biosensors. *Biosensors–Emerging Materials and Applications*. doi: 10.5772/17068.
48. Wenga, C. J., Chena, Y. L., Chiena, C. M., Hsua, S. C., Jhuoa, Y. S., Yeha, J. M., & Dai, C. F., (2013). Preparation of gold decorated SiO_2@polyaniline core-shell microspheres and application as a sensor for ascorbic acid. *Electrochim. Acta, 95*, 162–169.

49. Deiminiat, B., Razavipanah, I., Rounaghi, G. H., & Arbab-Zavar, M. H., (2017). A novel electrochemical imprinted sensor for acetylsalicylic acid based on polypyrrole, sol-gel and SiO_2@Au core-shell nanoparticles. *Sens. Actuators, B, 244*, 785–795.
50. Zhao, P., & Hao, J., (2015). 2, 6-Diaminopyridine-imprinted polymer and its potency to hair-dye assay using graphene/ionic liquid electrochemical sensor. *Biosens. Bioelectron., 64*, 277–284.
51. Cancar, H. D., Soylemez, S., Akpinar, Y., Kesik, M., Goker, S., Gunbas, G., Volkan, M., & Toppare, L., (2016). A novel acetylcholinesterase biosensor: Core-shell magnetic nanoparticles incorporating a conjugated polymer for the detection of organophosphorus pesticides. *ACS Appl. Mater. Interfaces, 8*, 8058–8067.
52. Lina, C. L., Panga, Y. S., Chaob, M. C., Chenb, B. C., Lina, H. P., Tangc, C. Y., & Lin, C. Y., (2008). Synthesis of SBA-16 and SBA-15 mesoporous silica crystals template with neutral block copolymer surfactants. *J. Phys. Chem. Solids, 69*, 415–419.
53. Hassan, H. M., Rahman, N. B. A., & Jalil, M. N., (2016). Mesoporous silica electrochemical sensors for the detection of ascorbic acid and uric acid. *Malaysian J. Anal. Sci., 20*, 351–357.
54. Nasir, T., Herzog, G., Hébrant, M., Despas, C., Liu, L., & Walcarius, A., (2018). Mesoporous silica thin films for improved electrochemical detection of paraquat. *ACS Sens., 3*, 484–493.
55. Yang, Y., & Yu, C., (2016). Advances in silica based nanoparticles for targeted cancer therapy *Nanomed-Nanotechnol., 12*, 317–332.
56. Castillo, R. R., Colilla, M., & Vallet-Regi, M., (2017). Advances in mesoporous silica-based nanocarriers for co-delivery and combination therapy against cancer. *Expert Opin. Drug Deliv., 14*, 229–243.
57. Agotegaray, M. A., & Lassalle, V. L., (2017). Silica: Chemical properties and biological features. *Silica-Coated Magnetic Nanopart.*, 27–37. doi: 10.1007/978-3-319-50158-1_3.
58. Wang, Y., Sun, L., Jiang, T., Zhang, J., Zhang, C., Sun, C., Deng, Y., et al., (2013). The investigation of MCM-48-type and MCM-41-type mesoporous silica as oral solid dispersion carriers for water-insoluble cilostazol. *Drug Dev. Ind. Pharm., 40*, 819–828.
59. Hu, Y., Zhi, Z., Wang, T., Jiang, T., & Wang, S., (2011). Incorporation of indomethacin nanoparticles into 3-D ordered macroporous silica for enhanced dissolution and reduced gastric irritancy. *Eur. J. Pharm. Biopharm., 79*, 544–551.
60. Bouchoucha, M., Côté, M. F., Gaudreault, R. C., Fortin, M. A., & Kleitz, F., (2016). Size-controlled functionalized mesoporous silica nanoparticles for tunable drug release and enhanced anti-tumoral activity. *Chem. Mater., 28*, 4243–4258.
61. Liu, R., Zhang, Y., Zhao, X., Agarwal, A., Mueller, L. J., & Feng, P., (2010). pH-responsive nanogated ensemble based on gold-capped mesoporous silica through an acid-labile acetal linker. *J. Am. Chem. Soc., 132*, 1500–1501.
62. Gan, Q., Lu, X., Yuan, Y., Qian, J., Zhou, H., Lu, X., Shi, J., & Liu, C., (2011). A magnetic, reversible pH-responsive nanogated ensemble based on Fe_3O_4 nanoparticles-capped mesoporous silica. *Biomaterials, 32*, 1932–1942.
63. Xu, C., Lin, Y., Wang, J., Wu, L., Wei, W., Ren, J., & Qu, X., (2013). Nanoceria-triggered synergetic drug release based on CeO_2-capped mesoporous silica host-guest interactions and switchable enzymatic activity and cellular effects of CeO_2. *Adv. Healthc. Mater., 2*, 1591–1599.

64. Feng, W., Zhou, X., He, C., Qiu, K., Nie, W., Chen, L., Wang, H., et al., (2013). Polyelectrolyte multilayer functionalized mesoporous silica nanoparticles for pH-responsive drug delivery: Layer thickness-dependent release profiles and biocompatibility. *J. Mater. Chem. B, 1*, 5886–5898.
65. Meng, H., Xue, M., Xia, T., Zhao, Y. L., Tamanoi, F., Stoddart, J. F., Zink, J. I., & Nel, A. E., (2010). Autonomous in vitro anticancer drug release from mesoporous silica nanoparticles by pH-sensitive nanovalves. *J. Am. Chem. Soc., 132*, 12690–12697.
66. Gao, Y., Yang, C., Liu, X., Ma, R., Kong, D., & Shi, L., (2012). A multifunctional nanocarrier based on annotated mesoporous silica for enhanced tumor-specific uptake and intracellular delivery. *Macromol. Biosci., 12*, 251–259.
67. Rim, H. P., Min, K. H., Lee, H. J., Jeong, S. Y., & Lee, S. C., (2011). pH-tunable calcium phosphate covered mesoporous silica nanocontainers for intracellular controlled release of guest drugs. *Angew. Chem. Int. Ed., 50*, 8853–8857.
68. Gisbert, M., Lozano, D., Vallet-Regí, M., & Manzano, M., (2017). Self-immolation polymers as novel pH-responsive gatekeepers for drug delivery. *RSC Adv., 7*, 132–136.
69. Zou, Z., He, D., He, X., Wang, K., Yang, X., Qing, Z., & Zhou, Q., (2013). Natural gelatin capped mesoporous silica nanoparticles for intracellular acid-triggered drug delivery. *Langmuir, 29*, 12804–12810.
70. Martínez-Carmona, M., Lozano, D., Colilla, M., & Vallet-Regí, M., (2017). Lectin-conjugated pH-responsive mesoporous silica nanoparticles for targeted bone cancer treatment. *Acta Biomater., 65*, 393–404.
71. Ma, X., Nguyen, K. T., Borah, P., Ang, C. Y., & Zhao, Y., (2012). Functional silica nanoparticles for redox-triggered drug/ssDNA co-delivery. *Adv. Healthc. Mater., 1*, 690–697.
72. Zhang, J., Niemelä, M., Westermarck, J., & Rosenholm, J. M., (2014). Mesoporous silica nanoparticles with redox responsive surface linkers for charge-reversible loading and release of short oligonucleotides. *Dalton Trans., 43*, 4115–4126.
73. Lai, C. Y., Trewyn, B. G., Jeftinija, D. M., Jeftinija, K., Xu, S., Jeftinija, S., & Lin, V. S. Y., (2003). A mesoporous silica nanosphere-based carrier system with chemically removable CdS nanoparticle caps for stimuli-responsive controlled release of neurotransmitters and drug molecules. *J. Am. Chem. Soc., 125*, 4451–4459.
74. Nadrah, P., Porta, F., Planinšek, O., Kros, A., & Gaberšček, M., (2013). Poly(propylene imine) dendrimer caps on mesoporous silica nanoparticles for redox-responsive release: Smaller is better. *Phys. Chem. Chem. Phys., 15*, 10740–10748.
75. Xua, J. H., Gao, F. P., Li, L. L., Ma, H. L., Fan, Y. S., Liu, W., Guo, S. S., et al., (2015). Gelatin–mesoporous silica nanoparticles as matrix metalloproteinases-degradable drug delivery systems in vivo. *Microporous Mesoporous Mater., 204*, 226–234.
76. Agostini, A., Mondragón, L., Coll, C., Aznar, E., Marcos, M. D., Martínez-Máñez, R., Sancenón, F., et al., (2012). Dual enzyme-triggered controlled release on capped nanometric silica mesoporous supports. *Chemistry Open, 1*, 17–20.
77. Van, R. S. H., Bölükbas, D. A., Argyo, C., Datz, S., Lindner, M., Eickelberg, O., Königshoff, M., et al., (2015). Protease-mediated release of chemotherapeutics from mesoporous silica nanoparticles to ex vivo human and mouse lung tumors. *ACS Nano, 9*, 2377–2389.

78. Singh, N., Karambelkar, A., Gu, L., Lin, K., Miller, J. S., Chen, C. S., Sailor, M. J., & Bhatia, S. N., (2011). Bioresponsive mesoporous silica nanoparticles for triggered drug release. *J. Am. Chem. Soc., 133*, 19582–19585.
79. Park, C., Kim, H., Kim, S., & Kim, C., (2009). Enzyme responsive nanocontainers with cyclodextrin gatekeepers and synergistic effects in release of guests. *J. Am. Chem. Soc., 131*, 16614–16615.
80. Mas, N., Arcos, D., Polo, L., Aznar, E., Sánchez-Salcedo, S., Sancenón, F., García, A., et al., (2014). Towards the development of 3D "gated scaffolds" for on-command delivery. *Small, 10*, 4859–4864.
81. Aznar, E., Villalonga, R., Giménez, C., Sancenón, F., Marcos, M. D., Martinez-Mañez, R., Díez, P., et al., (2013). Glucose-triggered release using enzyme-gated mesoporous silica nanoparticles. *Chem. Commun., 49*, 6391–6393.
82. Climent, E., Bernardos, A., Martínez-Máñez, R., Maquieira, A., Marcos, M. D., Pastor-Navarro, N., Puchades, R., et al., (2009). Controlled delivery systems using antibody-capped mesoporous nanocontainers. *J. Am. Chem. Soc., 131*, 14075–14080.
83. He, X., Zhao, Y., He, D., Wang, K., Xu, F., & Tang, J., (2012). ATP-responsive controlled release system using aptamer functionalized mesoporous silica nanoparticles. *Langmuir, 28*, 12909–12915.
84. Schferling, M., (2012). The art of fluorescence imaging with chemical sensors. *Angew. Chem. Int. Ed. Engl., 51*, 3532–3554.
85. Wang, J., Shah, Z. H., Zhang, S., & Lu, R., (2014). Silica-based nanocomposites via reverse microemulsions: Classifications, preparations and applications. *Nanoscale, 6*, 4418–4437.
86. Zhang, X., Zhang, X., Wang, S., Liu, M., Tao, L., & Wei, Y., (2013). Surfactant modification of aggregation-induced emission material as biocompatible nanoparticles: Facile preparation and cell imaging. *Nanoscale, 5*, 147–150.
87. Shi, M., Xia, L., Chen, Z., Lv, F., Zhu, H., Wei, F., Han, S., et al., (2017). Europium-doped mesoporous silica nanosphere as an immune-modulating osteogenesis/angiogenesis agent. *Biomaterials, 144*, 176–187.
88. Atabaev, T. S., Vu, H. H., Hwang, Y. H., & Kim, H. K., (2017). Ratiometric pH sensor based on fluorescent core-shell nanoparticles. *J. Nanosci. Nanotechnol., 17*, 8313–8316.
89. Jiao, L., Song, F., Zhang, B., Ning, H., Cui, J., & Peng, X., (2017). Improving brightness and photostability of NIR fluorescent silica nanoparticles through rational fine-tuning of the covalent encapsulation methods. *J. Mater. Chem. B, 5*, 5278–5283.
90. Geng, J., Goh, C. C., Qin, W., Liu, R., Tomczak, N., Ng, L. G., Tang, B. Z., & Liu, B., (2015). Silica shelled and block copolymer encapsulated red-emissive AIE nanoparticles with 50% quantum yield for two-photon excited vascular imaging. *Chem. Commun., 51*, 13416–13419.
91. Kumar, R., Roy, I., Ohulchanskyy, T. Y., Goswami, L. N., Bonoiu, A. C., Bergey, E. J., Tramposch, K. M., et al., (2008). Covalently dye-linked, surface-controlled and bioconjugated organically modified silica nanoparticles as targeted probes for optical imaging. *ACS Nano, 2*, 449–456.
92. Francis, B., Neuhaus, B., Reddy, M. L. P., Epple, M., & Janiak, C., (2017). Amine-functionalized silica nanoparticles incorporating covalently linked visible-light excitable Eu^{3+}-complexes: Synthesis, characterization and cell uptake studies. *Eur. J. Inorg. Chem., 25*, 3205–3213.

93. Li, Y., Jiao, J., Yan, P., Liu, L., Wang, J., Wang, Y., Huang, L., et al., (2018). Synthesis and tunable photoresponse for core-shell structured $NaGdF_4$: Yb,Er@SiO_2@ Eu(TTA)$_3$Phen nanocomplexes. *Scr. Mater., 152*, 1–5.
94. Cho, W., Lee, H. J., Choi, S., Kim, Y., & Oh, M., (2014). Highly effective heterogeneous chemosensors of luminescent silica@coordination polymer core-shell microstructures for metal ion sensing. *Sci. Rep., 4*. doi: 10.1038/srep06518.
95. Jin, Y., Lohstreter, S., Pierce, D. T., Parisien, J., Wu, M., Hall, C., & Zhao, J. X., (2008). Silica nanoparticles with continuously tunable sizes: Synthesis and size effects on cellular contrast imaging. *Chem. Mater., 20*, 4411–4419.
96. Mirenda, M., Levi, V., Bossi, M. L., Bruno, L., Bordoni, A. V., Regazzoni, A. E., & Wolosiuk, A., (2013). Temperature response of luminescent tris(bipyridine) ruthenium(II)-doped silica nanoparticles. *J. Colloid Interface Sci., 392*, 96–101.
97. Savchuck, O. A., Carvajal, J. J., Cascales, C., & Diaz, F., (2016). Benefits of silica core–shell structures on the temperature sensing properties of Er,Yb: $GdVO_4$ up-conversion nanoparticles. *ACS Appl. Mater. Interfaces, 8*, 7266–7273.
98. Wang, N., Koh, S., Jeong, B. G., Lee, D., Kim, W. D., Park, K., Nam, M. K., et al., (2017). Highly luminescent silica-coated CdS/CdSe/CdS nanoparticles with strong chemical robustness and excellent thermal stability. *Nanotechnology,* 28. doi: 10.1088/1361-6528/aa6828.
99. Ji, B., Giovanelli, E., Habert, B., Spinicelli, P., Nasilowski, M., Xu, X., Lequeux, N., et al., (2015). non-blinking quantum dot with a plasmonic nanoshell resonator. *Nat. Nanotecgnol., 10*, 170–175.
100. Chang, K., Men, X., Chen, H., Liu, Z., Yin, S., Qin, W., Yuan, Z., & Wu, C., (2015). Silica-encapsulated semiconductor polymer dots as stable phosphors for white light-emitting diodes. *J. Mater. Chem. C., 3*, 7281–7285.

CHAPTER 4

Tamarind Kernel Powder, Its Derivatives, and Their Modification Through Grafting: An Overview

J. H. TRIVEDI,[1] AGEETHA VANAAMUDAN,[2] and H. C. TRIVEDI[2]

[1]*Post Graduate, Department of Chemistry, Sardar Patel University, Vallabh Vidyanagar–388120, Gujarat, India, E-mail: drjignesh2575@yahoo.co.in*

[2]*Department of Chemistry, Faculty of Applied Sciences, Parul University, Limda, Vadodara–391760, Gujarat, India*

ABSTRACT

Tamarind Kernel Powder (TKP), a highly branched neutral, non-ionic polysaccharide with considerably high molecular weight, is derived from the seeds of the evergreen tamarind tree (Tamarindus Indica L.). This chapter briefly describes the chemical structure, industrial importance as well as chemical and physical modifications of TKP. Besides, it also provides a comprehensive literature survey review on the modification of TKP and its derivatives through grafting, which may be of immense utility for carrying out further progress in research with a view to exploring potential applications in terms of food, pharmaceuticals, and industrial benefits.

Advances in Diverse Applications of Polymer Composites: Synthesis, Application, and Characterization.
Suji Mary Zachariah, Yang Weimin, Maciej Jaroszewski, & Sabu Thomas (Eds.)

4.1 INTRODUCTION

Tamarind Kernel Powder (TKP) or tamarind seed powder (TSP) or tamarind seed polysaccharide (TSP) or Tamarind Gum (TG) is derived from the seed of the evergreen tamarind tree. The tamarind tree is usually known as Imli or Amli or Chinch, and its botanical name is Tamarindus Indica Linn. The major countries involved in the cultivation of tamarind trees include India, Bangladesh, East Pakistan, Sri Lanka, Thailand, Indonesia, Africa, Mexico, and South America. In India, it is particularly abundant in the states of Madhya Pradesh, Karnataka, West Bengal, Andhra Pradesh, Bihar, and Tamil Nadu. However, today, India is the largest producer of tamarind products in the world [1].

Amongst various parts of the tamarind tree like root, body, fruit, flowers, and leaves, the most valuable and commonly used part is the tamarind fruit (known as pod also) which consists mainly of pulp and seeds (Figure 4.1). All parts of the tamarind, being a great variety of bioactive substances have a high nutritional, industrial, and economic importance [2, 3]. The tamarind seed consists of the seed coat or Testa or husk (rich brown color, 20–30%) and the white kernel or endosperm (70–75%) [4, 5].

The tamarind seeds are of great industrial importance as they are used in the manufacture of TKP. TKP is prepared by decorticating the seeds and pulverizing the creamy white kernels or endosperms [6, 7] (Figure 4.1). On the other hand, to prepare TSP, tamarind seeds are directly milled several times and finally sieved to the required mesh size to obtain a powder [7] (Figure 4.1).

Tamarind seeds and kernels or endosperms are high in protein content, so they can be used as a cheaper source of protein in the areas of the world where protein malnutrition is a widespread problem. The seed coat is rich in fiber (20%) and tannins (20%) as well. Several workers have determined the chemical composition and nutritive value of tamarind seeds and kernels [8, 9]. The tamarind fruit pulp contains carbohydrates, proteins, organic acids (malic, tartaric, citric, succinic), vitamins, and minerals. According to the World Health Organization, tamarind can be regarded as a source of all essential amino acids (with the exception of tryptophan) and as a food as it is rich in nutrients, fibers, and phenolic compounds with antioxidant action.

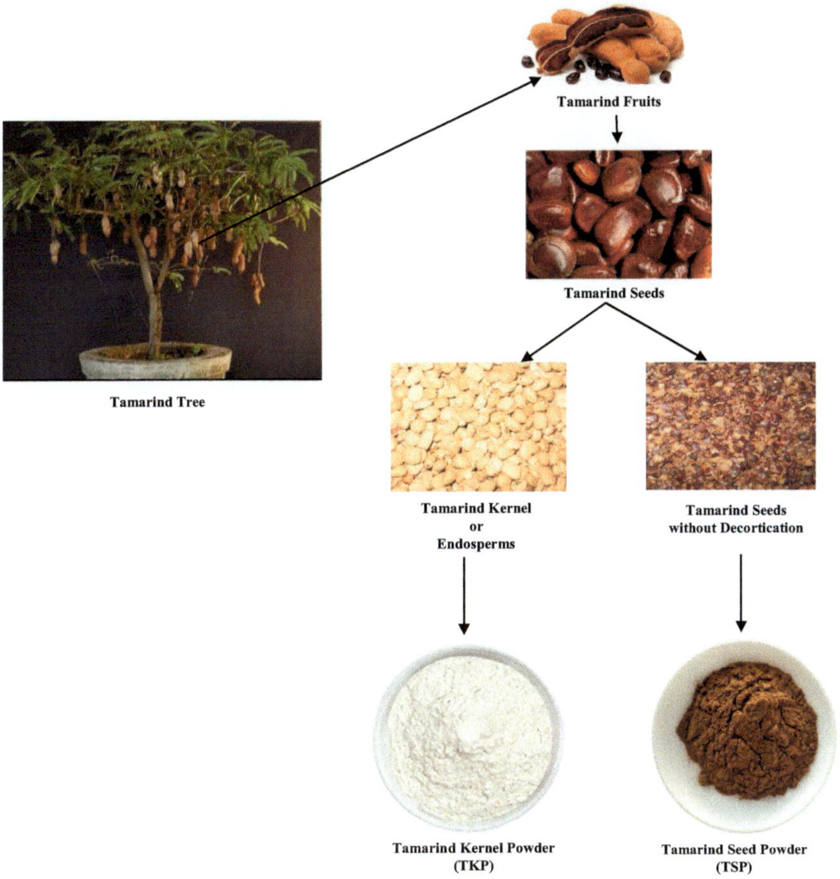

FIGURE 4.1 Tamarind tree: TKP and TSP.

4.2 CHEMICAL STRUCTURE

TKP–a highly branched neutral, non-ionic polysaccharide with considerably high molecular weight, is composed of D-galactose, D-xylose, and D-glucose in the molar ratio of 1:2:3 [10]. It consists of a main chain of β-D-(1→4) linked glucopyranosyl units and a side chain consisting of a single xylopyranosyl unit is attached to every second, third, and fourth D-glucopyranosyl unit through an α-D-(1→6) linkage. One D-galactopyranosyl unit is attached to one of the xylopyranosyl units through a β-D-(1→2) linkage [10–12] (Figure 4.2(a) and (b)). Thus, TKP consists of a cellulose-type backbone that carries xylopyranosyl and galactopyranosyl units.

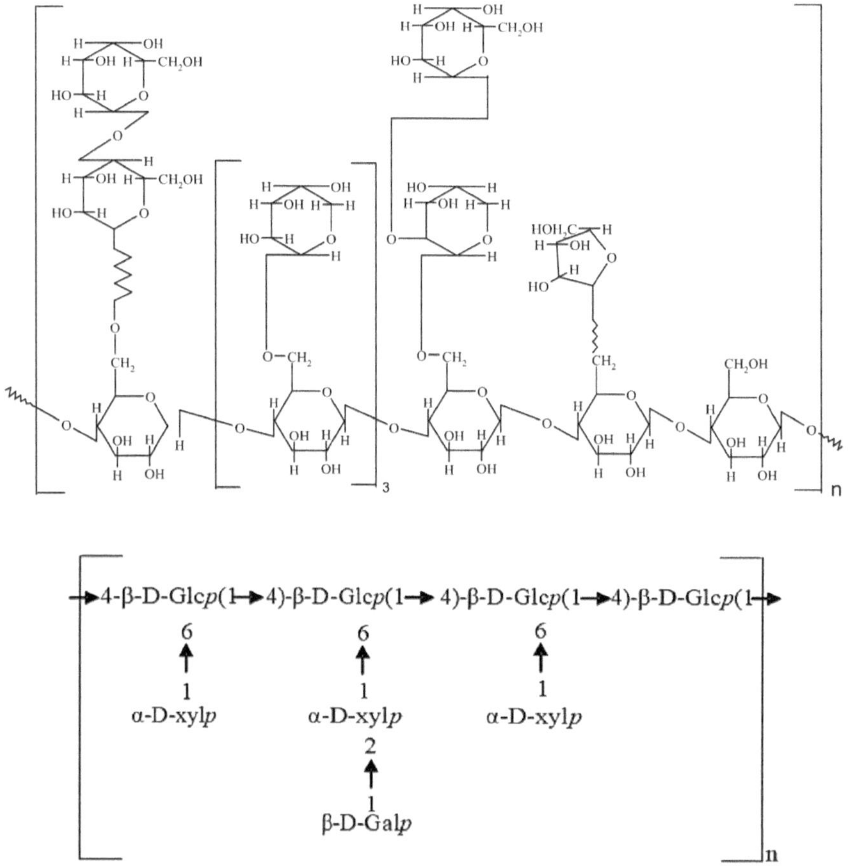

FIGURE 4.2 Structure of tamarind kernel powder (TKP).

Source: a) Reprinted from Ref. [66].

4.3 INDUSTRIAL IMPORTANCE

TKP, an industrially important polysaccharide, is known for its wide range of Food and Non-food applications in a number of industries [13, 14].

In food industries, TKP is used as a gelling, thickening, and bulking agent as well as an emulsifier, stabilizer, and crystallization inhibitor for ice cream, frozen food, etc. It is also used as a fruit preservative and also in jelly preparations.

The non-food applications of TKP mainly include its uses in the textile, plywood, paper, and pharmaceutical industries. TKP is used as a thickening, sizing, and printing agent in the jute and textile industries. In plywood industries, it is used as an extender for urea-formaldehyde resin adhesives. Paper coated with xyloglucan (XG) in a spray application showed improved mechanical strength, and therefore in the paper industry, TKP can substitute starch in many adhesive applications and as a binder in particle board, corrugated board, and books.

TKP is widely used in pharmaceuticals as a binder in the manufacture of pills and tablets, as an excipient in making greaseless ointments, and as a gelatinizing agent in the preparation of colloidal iodine jelly, etc. However, recent reviews [15–18] have highlighted the utility of TKP and its derivatives for pharmaceutical applications. Kuru [19] has provided a valuable review furnishing comprehensive information on the medicinal utilization of Tamarind. In addition, TKP is used as a thickener in explosives, as a soil stabilizer, and in cosmetic preparations like shaving creams and moisturizing creams [13].

4.4 TKP DERIVATIVES

In India, TKP though it is the cheapest gum, it is having several drawbacks [11], such as unpleasant odor due to fat (7%), dull color, presence of water-insoluble ingredients, and low solubility in cold water. In order to overcome these drawbacks, molecular modification of TKP is required to be carried out by employing a variety of chemical and physical methods (Table 4.1). This will help in enhancing the quality and acceptability of TKP and its products for various industrial applications.

TABLE 4.1 TKP Derivatives

Method of Modification	Derivative(s)	References
Chemical method	Acetylation	[20]
	Thiolation	[21, 22]
	Carboxylation, sulfonation, and alkyl amination	[23]
	Allylation	[24]
	Cyanoethylation	[25]
	Degalactosylation	[26]
	Cross-linking with epichlorohydrin	[27, 28]
	Cross-linking with glutaraldehyde	[29, 30]
	Carboxymethylation	[31–34]
Physical method	Composite hydrogels of CMTKP and PVA	[35]

4.4.1 CHEMICAL MODIFICATION

Chemical modification is a common method that can change the structure of TKP by introducing substituent groups and thereby strengthen its original biological activities as well as create new functional bioactivities. Chemical modification methods mainly include carboxymethylation, cyanoethylation, carboxylation, sulfonation, alkylamination, acetylation, thiolation, allylation, etc.

4.4.1.1 ACETYLATION

The hexa, duodeca, and hexadeca acetyl derivatives of tamarind seed jellose were prepared by treatment with acetic anhydride under different conditions [20].

4.4.1.2 THIOLATION

Thiol functionalization of TSP was accomplished by esterification of its hydroxyl groups with thioglycolic acid. Thiol-functionalization was confirmed by using FTIR, DSC, XRD, and SEM techniques. Mucoadhesive characteristics of TSP were comparatively evaluated with thiolated-TSP by conducting tensile tests. Thiolated TSP was further explored as mucoadhesive agent by formulating Carbopol-based metronidazole gels. These gels were compared with the marketed formulation with regard to the mechanical characterization, viscosity, mucoadhesive strength, and in vitro drug release [21].

Mahajan et al. [22] have carried out thiolation of XG and thiolated XG was characterized by NMR, DSC, and XRD analysis. Thiolated XG was further successfully explored for mucoadhesive applications by developing in situ gel system employing ondansetron as a model drug.

4.4.1.3 CARBOXYLATION, SULFONATION, ALKYLAMINATION

Lang et al. [23] have prepared carboxylated, sulfated, and alkylaminated derivatives of TSP. The nature of the extent of substitution has been characterized by potentiometric titration, infrared, and ^{1}H and ^{13}C NMR (nuclear magnetic resonance) spectroscopy. The solution properties of the carboxylated and sulfated derivatives were also examined.

4.4.1.4 ALLYLATION

A novel method for the introduction of alkyl groups to galactose-containing polysaccharides viz. galactoglucomannan (GGM), guar galactomannan (GM), and tamarind (galacto) XG using enzymatic oxidation combined with indium mediated allyation was developed. The formation of the alkylated or propargylated product was identified by the Matrix-assisted laser desorption/ionization time-of-flight mass spectrometry. All polysaccharide products were isolated and further characterized by GC-MS (gas chromatography-mass spectrometry) or NMR spectroscopy [24].

4.4.1.5 CYANOETHYLATION

Goyal et al. [25] have carried out cyanoethylation of TKP with acrylonitrile (AN) in the presence of sodium hydroxide as a catalyst. The optimum reaction conditions for cyanoethylation of TKP were established by varying various reaction conditions, including concentrations of sodium hydroxide and AN, gum-liquor ratio as well as reaction temperature, and time. The degree of substitution ($\overline{DS}$), reaction efficiency, and total extent of etherification were also determined. Rheological studies showed the non-Newtonian pseudoplastic nature of cyanoethyl-TKP.

4.4.1.6 DEGALACTOSYLATION

The degalactosylation of XG was monitored in real-time using time-dependent static light scattering (SLS), viscometry, and HPSEC-MALLS (high-performance size-exclusion chromatography-Multiangle laser light scattering). The results confirmed that aggregation is the main phenomenon responsible for gelling of degalactosylated XG and that the conformation was practically unaffected by the enzymatic treatment [26].

4.4.1.7 CROSS-LINKING WITH EPICHLOROHYDRIN

The sustained release behavior of both water-soluble (acetaminophen, caffeine, theophylline, and salicylic acid) and water-insoluble (indomethacin) drugs from TSP was studied. TSP was partially cross-linked with epichlorohydrin, and it was observed that for the water-soluble drugs, the release amount could be controlled by partially cross-linking the matrix. The extent of release was found to be varied by controlling the degree of cross-linking [27].

Babu et al. [28] have extracted XG from the kernel of tamarind, and its identification was performed using LC-MS/MS (liquid chromatography-mass spectrometer), FTIR, and NMR studies. XG was cross-linked with epichlorohydrin, and the obtained gel polymer was tested as a wound-healing agent and also as a slow drug-releasing scaffold to help wound healing.

4.4.1.8 CROSS-LINKING WITH GLUTARALDEHYDE

Shaw et al. [29] have synthesized gelation-tamarind gum (TG)/carboxymethyl tamarind (CMT) gum-based phase-separated hydrogels using glutaraldehyde as cross-linking agent. The hydrogels were thoroughly characterized using a bright-field microscope, FTIR spectroscope, differential scanning calorimeter, mechanical tester, and impedance analyzer. The mucoadhesivity, biocompatibility, and swelling properties of the hydrogel were also evaluated. To understand the ability of the synthesized hydrogels as vehicles for controlled release, the hydrogels were loaded with ciprofloxacin (fluoroquinolone antibiotic). The drug release kinetics and the antimicrobial activity of the drug-loaded hydrogels were also studied in depth.

Ajovalasit et al. [30] have synthesized thin XG-based hydrogel films and characterized them for wound dressing. The influence of Polyvinyl alcohol (PVA), chemical cross-linking induced by glutaraldehyde, and glycerol content on the XG-based hydrogel film molecular structure were investigated by rheological and thermal analyzes. The change in hydrogel morphology, water retention, and swelling behavior in water and physiologic buffer were also studied. In order to assess the cytotoxicity of the films, Biological in vitro tests were also carried out.

4.4.1.9 CARBOXYMETHYLATION

Prabhanjan [31] has prepared eight batches of sodium salt of carboxymethyl derivatives of TKP, at different levels of substitution, in isopropanol medium. All the nine batches of carboxymethyl TKP were characterized by degree of substitution ($\overline{DS}$), intrinsic viscosity and elution profile (GPC: gel permeation chromatography). Solution properties of the derivatives as a function $\overline{DS}$ were found to be influenced by the substituent groups as well as by the alkali used for derivatization.

Carboxymethylation of TKP was carried out with monochloroacetic acid in the presence of alkali as a catalyst under heterogeneous conditions. The carboxymethyl reaction conditions were optimized with respect to various reaction variables viz. concentrations of sodium hydroxide and monochloroacetic acid, solvent ratio, reaction time, and temperature [32].

Pal et al. [33] have prepared carboxymethylated tamarind (CMT) sample by reacting TKP with sodium salt of monochloroacetic acid in the presence of sodium hydroxide. CMT was characterized by using the spectroscopic (FTIR, ^{13}CNMR), thermal (TGA and DTA), elemental analysis (C, H, N, and O), viscosity, and SLS techniques. The CMT sample was also evaluated by in vitro drug release studies.

Ponnikornkit et al. [34] have carried out carboxymethylation of tamarind gum at different mole ratios of sodium hydroxide to monochloroacetic acid using methanol as solvent in order to improve the properties of tamarind gum and extend its application in the area of pharmaceutical science. The swelling behavior of all the samples of carboxymethylated tamarind gum was also compared with the crude one and the results showed that the carboxymethylated tamarind gum ($\overline{DS}$ = 0.1711) could be better used in comparison with other samples for further pharmaceutical applications as disintegrant, diluent, and drug release controlling agent.

4.4.2 PHYSICAL MODIFICATION

The Freeze-thaw cycling method is one of the promising methods for physical modification of natural polymers. According to this method, physical cryogels based on natural polymers are formed on subjecting aqueous dispersion of natural polymers to freeze-thaw cycling. The cryogels which are formed are biocompatible, biodegradable, and eco-friendly.

Meenakshi and Ahuja [35] have prepared composite hydrogels of carboxymethyl tamarind kernel polysaccharide (CMTKP) and polyvinyl alcohol (PVA) using freeze-thaw cyclic method and evaluated them for studying the release behavior of a model drug (Metronidazole). The preparation of CMTKP-PVA composite cryogels was optimized using a three-factor, three-level central composite experimental design employing the concentrations of CMTKP, PVA, and the number of freeze-thaw cycles as the independent variables. The optimized composite cryogels were characterized by FTIR, XRD, SEM, and TGA studies.

4.5 MODIFICATION OF TKP AND ITS DERIVATIVES THROUGH GRAFTING

Chemical modification of polymers with the aim of imparting specific desirable properties is one of the main directions of the development of modern macromolecular chemistry. In recent years, the chemical modification of natural, renewable polymers by grafting has received considerable attention as it imparts useful desirable properties (such as thermal stability, dye-ability, water repellency, flame resistance and resistance towards acid-base attack, etc.), to the backbone polymer without affecting the architecture of the original ones [36]. Modification of natural polymers, especially polysaccharides through graft copolymerization, may be initiated by chemical treatment, photo-irradiation, high-energy radiation technique, etc. [36, 37]. The synthesis of polysaccharide-based graft copolymers has made a significant contribution toward improved industrial and biomedical applications.

Recently, considerable interest has also been provoked in the area of polysaccharide-based superabsorbent hydrogels because of their non-toxicity, hydrophilicity, biocompatibility, and biodegradability [38].

Due to their excellent characteristics, superabsorbent hydrogels exhibit potential applications in many fields such as hygienic products, agriculture, and horticulture, pharmaceutics, and medicine, wastewater treatment, and metal-ion removal [39, 40].

Even though TKP and its derivatives find a wide range of industrial applications, they also suffer from some drawbacks like biodegradability [12], which limits their uses considerably. However, these drawbacks can be improved through the grafting of vinyl monomers onto them, thereby imparting new properties to the polymer backbone.

A comprehensive literature survey reveals that some literature reports are available on the modification of TKP and its derivatives through grafting employing mainly chemical, low-energy radiation, and microwave-assisted methods (Table 4.2). However, there are very few reports on the synthesis, characterization, and evaluation of the superabsorbent hydrogels based on TKP and its modified products.

Mishra and Bajpai [41] have carried out grafting of acrylamide (AAm) onto a water-soluble food-grade polysaccharide, Tamarind mucilage (TAM), using the CAN/HNO_3 redox initiator system. The influence of reaction variables like concentrations of monomer and initiator as well as reaction time and the temperature was also studied in terms of grafting efficiency (%GE) and percent of grafting (%G).

Mishra et al. [42] have synthesized the graft copolymer, TAM-g-PAM, by grafting AAm onto TAM by a radical polymerization method in an aqueous system, using a ceric ion/nitric acid redox initiator. The authors have also assessed the application of Tamarindus indica seed mucilage (TAM) and its graft copolymer (TAM-g-PAM) for the removal of various types of dyes from model textile wastewater containing azo, basic, and reactive dyes. The variables studied were the flocculant dose, contact time, and pH. TAM-g-PAM showed better flocculation efficiency for dye removal than pure mucilage. These flocculants performed better for the removal of azo and reactive dyes than that of basic dyes.

Goyal et al. [43] have carried out graft copolymerization of AAm onto TKP in an aqueous medium using ceric ammonium nitrate (CAN)-nitric acid initiator system. The reaction conditions were optimized for grafting by varying various reaction conditions, and the maximum percentage of grafting and percentage of grafting efficiency were achieved to be 231.45% and 93.66%, respectively.

Mishra et al. [44] have extracted XG from tamarind seed mucilage and grafting of AN onto XG was carried out using the CAN/HNO_3 redox initiator system as well as microwave (MW) irradiation. The influence of different reaction conditions, including concentrations of monomer and initiator as well as reaction time, temperature, and MW power on the percent grafting was studied. The grafting was ascertained by FTIR, DSC, and SEM techniques.

Singh et al. [45] have successfully grafted AN onto tamarind seed gum (TSG) using persulfate/ascorbic acid redox initiator. The grafting conditions were optimized by varying various reaction parameters and the maximum

TABLE 4.2 Modification of TKP and Its Derivatives Through Grafting

Polysaccharide	Monomer	Method of Initiation	Application(s)	References
TAM	AAm	Chemical	• Drug delivery • Flocculant	[41]
TAM	AAm	Chemical	• Flocculant	[42]
TKP	AAm	Chemical	• Flocculant	[43]
XG	AN	Chemical and microwave-irradiation	• Drug delivery system	[44]
TSG	AN	Chemical	• Adsorbent	[45]
TKP	AN	Chemical	• Adsorbent	[46]
CMT	AAm	Chemical	• Flocculant	[47]
TKP	AAm	Chemical, microwave-irradiation, microwave-assisted	• Flocculant	[48]
TKP	AAm	Chemical and partial alkaline hydrolysis of the graft copolymer	• Flocculant	[49]
TSP	MMA	Chemical	• Drug delivery	[50]
XG	MMA	Chemical	• Drug delivery	[51]
TKP	AAm	Microwave-assisted	• Drug delivery	[52]
Na-PCMTKP	AN	Chemical	• Personal health care product	[53]
TSP	NVP	Microwave-assisted	• Drug delivery	[54]
Na-PCMTKP	AN	Chemical and alkaline hydrolysis of the graft copolymer	• Personal health care product • Agriculture and horticulture field product	[55]
CMTKP	AN	Microwave-assisted	• pH-responsive hydrogels	[56]

TABLE 4.2 *(Continued)*

Polysaccharide	Monomer	Method of Initiation	Application(s)	References
TKP	EA	Chemical	• Environmentally friendly polymeric material	[57]
TKP	MAAc	SI-ATRP	• Adsorbent for toxic dyes	[58]
TSP	AAm	Microwave Irradiation	• Drug delivery	[59]
TG	MMA	ATRP	• Adsorbent for toxic dyes	[60]
Na-PCMTKP	MA	Photo-irradiation	• Adsorption for metal ion sorption	[61]
CMT	HEMA	Chemical	• Skin tissue engineering and skin regeneration	[62]
TSG	AAm	Microwave-assisted	• Flocculant	[63]
Na-PCMTKP	MMA	Photo-irradiation	• Metal adsorbent	[64]
	EMA		• Excipient for direct compression matrix tablets	
Na-PCMTKP	AN	Photo-irradiation	• Personal health care products	[65]

values of %grafting and %grafting efficiency were found to be 305% and 75%, respectively. The synthesized graft copolymer was characterized using FTIR, XRD, and SEM techniques. The gel-forming ability, water/saline retention and the shelf life of the grafted gum solutions were also studied.

Goyal et al. [46] have carried out graft copolymerization of AN onto TKP using CAN/HNO_3 initiating system. The reaction conditions were optimized to afford the maximum percentage of grafting. The maximum values of %G and %GE were found to be 86% and 64%, respectively. The grafted products were characterized by FTIR, SEM, and thermal analysis.

Sen and Pal [47] have developed a novel polymeric flocculant by grafting polyacrylamide (PAM) onto CMT. The grafted copolymers (CMT-g-PAM: carboxymethyl tamarind-graft-polyacrylamide) were characterized by viscosity measurements, spectroscopic (FTIR and ^{13}C-NMR) techniques, and elemental analysis. The flocculation studies were carried out using a turbidity test as well as a settling test. The performance of the synthesized flocculant was compared with other commercially available flocculants, and the performance of the synthesized flocculant (CMT-g-PAM) was found to be best at the laboratory scale.

Ghosh et al. [48] have prepared novel biodegradable polymeric flocculants by conventional redox grafting, microwave-initiated, and microwave-assisted grafting of AAm onto tamarind kernel polysaccharide (TKP). The synthesized graft copolymers were characterized by different techniques like viscometry, elemental analysis, molecular weight determination using SLS analysis, and NMR spectroscopy. The flocculation efficiency of the grafted products was evaluated by measurement of turbidity and reduction of total pollutant load of different wastewaters.

Ghosh et al. [49] have synthesized a high-performance anionic flocculant by partial alkaline hydrolysis of polyacrylamide grafted tamarind kernel polysaccharide (TKP-g-PAM). The products were characterized by using viscosity measurements, elemental analysis, FTIR spectroscopy, and TGA analysis. The flocculation characteristics of the hydrolyzed and unhydrolyzed grafted products were studied in kaolin suspension as well as municipal sewage wastewater. The performance of the hydrolyzed product was found to be better than the unhydrolyzed grafted TKP.

Shailaja et al. [50] have grafted methyl methacrylate (MMA) onto TSP using potassium persulfate and ascorbic acid redox pair. In order to optimize the grafting process, Taguchi L9 design was applied. The grafted tamarind seed polysaccharide (GTS) was characterized by using

FTIR, XRD, and SEM techniques. A flexible buccal patch for delivery of metoprolol succinate drug was developed using TSP and grafted TSP (GTSP). The results of in *vitro* and *ex vivo* studies showed that the TSP and GTSP could be successfully used to develop buccal patches with sustained delivery of metoprolol succinate for 12 hrs.

Mishra and Malhotra [51] have reported grafting of polymethyl methacrylate (PMMA) with XG using CAN/HNO_3 redox initiator system. The grafting was confirmed by using FTIR and NMR spectroscopic techniques as well as thermal (TGA and DSC) and SEM measurements.

Ghosh and Pal [52] have developed a novel hydrogel based on the synthesized graft copolymer viz. TKP-g-PAM. The graft copolymer was synthesized by grafting of polyacrylamide onto tamarind kernel polysaccharide (TKP) using the microwave-assisted method. The synthesized hydrogels were used as matrix, to study the controlled release behavior of a model drug aspirin by following the USP drug dissolution (basket type) method at various pH environments. It was observed that the rate of release of the enclosed drug from the matrix was low in acidic environments and much higher in neutral and alkaline environments.

Trivedi et al. [53] have modified sodium salt of partially carboxymethylated TKP (Na-PCMTKP, $\overline{DS}$ = 0.15) through grafting with AN using CAN as a redox initiator in an aqueous medium. The optimum reaction conditions for affording the maximum percentage of grafting were established by successively varying reaction conditions such as concentrations of nitric acid, CAN, monomer (AN), as well as reaction time, temperature, and amount of substrate. The influence of these reaction conditions on the grafting yields was discussed. The experimental results were also analyzed in light of the proposed kinetic scheme, and the results were found to be in very good agreement with the proposed scheme. The optimized graft copolymer was saponified using 0.7N sodium hydroxide solution at 90–95C to yield the superabsorbent hydrogel, and the swelling behavior of the hydrogel was studied. FTIR, TGA, and SEM techniques were used to characterize the products.

Ahuja et al. [54] have prepared graft copolymer of N-vinyl-2-pyrrolidone and TSP using microwave-assisted method. The optimization of the graft copolymerization was carried out using response surface methodology employing microwave power and exposure time as the independent variables and grafting efficiency as the response variables. The synthesized graft copolymer was further evaluated for mucoadhesive application by formulating the buccal patch using metronidazole as a model drug. The buccal patches were also evaluated for weight variation, friability, disc thickness, assay, *in vitro* release, and *ex vivo* bioadhesion time.

Trivedi [55] has prepared a novel superabsorbent hydrogel, H-Na-PCMTKP-g-PAN, followed by alkaline hydrolysis (0.7N KOH solution at 90–95°C) of the graft copolymer of sodium salt of partially carboxymethylated TKP containing polyacrylonitrile (Na-PCMTKP-g-PAN, %G = 413.76 and %GE = 96.48). The mechanism of the conversion of the nitrile groups of Na-PCMTP-g-PAN into a mixture of hydrophilic carboxamide and carboxylate groups during alkaline hydrolysis, followed by in situ cross-linking of the grafted PAN chains was proposed (Figure 4.3). The products were characterized spectroscopically (FTIR) and morphologically (SEM). The swelling behavior of the unreported superabsorbent hydrogel, H-Na-PCMTKP-g-PAN, was studied by carrying out its absorbency measurements in low-conductivity water, 0.15 M salt (NaCl, $CaCl_2$ and $AlCl_3$) solutions and simulated urine (SU) at different timings. The underlying principle behind the ionic dependence of swelling of the hydrogel (H-Na-PCMTKP-g-PAN) was well explained by the Donnan Equilibrium theory (Figure 4.4).

The superabsorbent hydrogel exhibited a high capability of water absorption (absorbency in water, 242.05 g/g gel; absorbency in 0.15 M NaCl, $CaCl_2$ (calcium chloride), $AlCl_3$, and SU solutions to be 65.85 g/g gel, 62.43 g/g gel, 41.88 g/g gel and 63.51 g/g gel, respectively). The results regarding the absorbency measurements of the superabsorbent hydrogel in different swelling media were explained on the basis of the "charge screening" effect and the "ionic cross-linking" phenomenon. The experimental data clearly suggested that the swelling process of the hydrogel obeys second-order kinetics in different swelling media. The values of the swelling characteristics, as well as equilibrium water contents of the hydrogel in different swelling media, were also reported. The obtained promising results lead to conclude that the prepared hydrogel may be suitable for utilization as diaper as well as adsorbent material.

Meenakshi et al. [56] have carried out a microwave-assisted synthesis of graft copolymers of carboxymethyl tamarind seed polysaccharide (CMTKP) and polyacrylonitrile (PAN). The synthesis of CMTKP-g-PAN (carboxymethyl tamarind kernel powder-graft-polyacrylonitrile) was optimized by varying various reaction conditions viz concentrations of AN and ammonium persulfate (APS), microwave irradiated time, and microwave power. The formation of the graft copolymer was confirmed by FTIR, TGA, DSC, XRD, and SEM studies. The synthesized graft copolymer (CMTKP-g-PAN) showed pH-dependent swelling behavior.

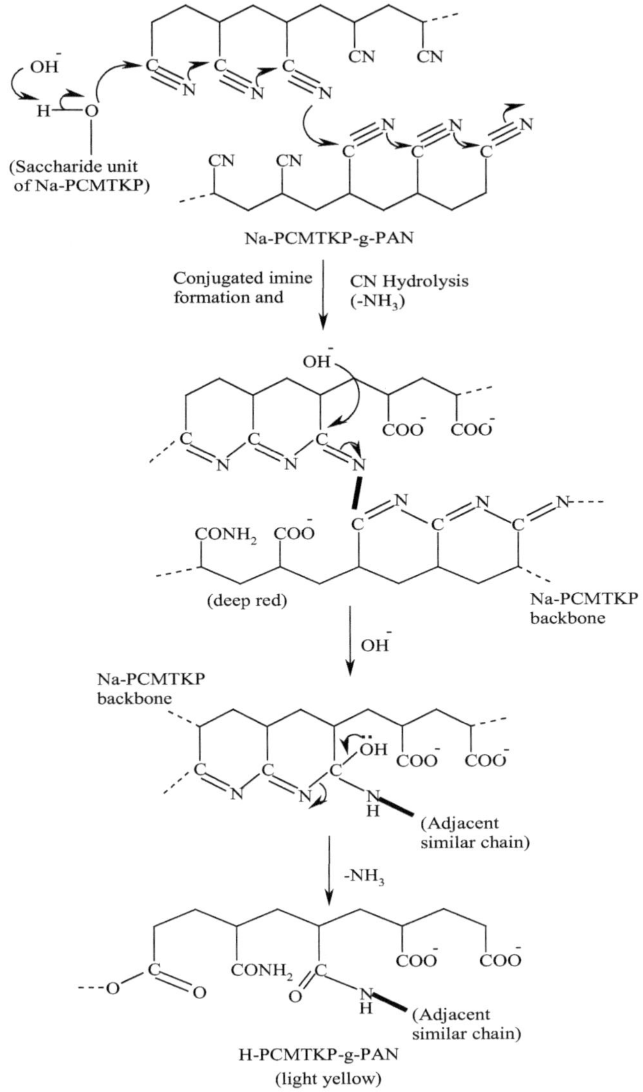

FIGURE 4.3 Mechanism of cross-linking during conversion of nitrile groups of Na-PCMTKP-g-PAN into carboxamide and potassium carboxylate groups for the formation of superabsorbent hydrogel, H-Na-PCMTKP-g-PAN.

Source: a) Reprinted from Ref. [66].

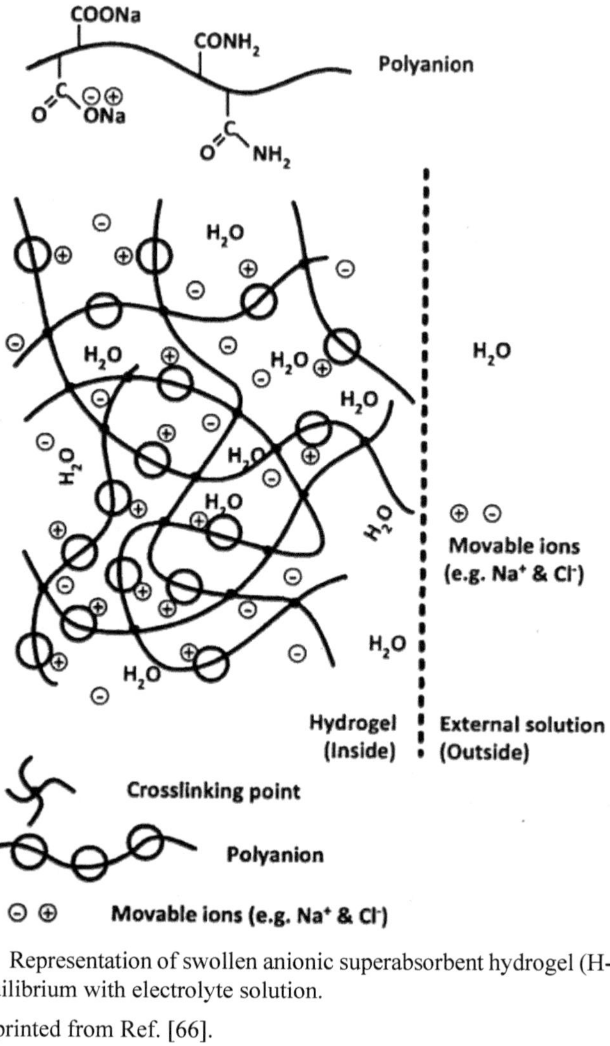

FIGURE 4.4 Representation of swollen anionic superabsorbent hydrogel (H-Na-PCMTP-g-PAN) in equilibrium with electrolyte solution.

Source: a) Reprinted from Ref. [66].

Del Real et al. [57] have grafted ethyl acrylate (EA) successfully onto TKP using azobisisobutyronitrile (AIBN) as the initiator. The synthesized graft copolymer was characterized by FTIR, ^{1}H-NMR, TGA, DSC, and SEM techniques. The biodegradability of the grafted copolymer was evaluated by subjecting the graft copolymer to cultures of the bacterial strain *Alicycliphilus* sp. BQ1. The newly synthesized graft copolymer manifested a steady process of biodegradation, and this was proved by SEM and IR studies.

Pal and Pal [58] have synthesized a novel copolymer derived from TKP and polymethacrylic acid (TKP-g-PMAAc) through surface-initiated atom transfer radical polymerization (SI-ATRP). The copolymer was characterized by FTIR and ^{1}H-NMR spectroscopy, elemental analysis, TGA, GPC analysis, and SEM studies. The synthesized graft copolymer exhibited excellent methylene blue (MB) and Erichrome black T (EBT) adsorption capacity with a rapid sorption rate in the basic media and in the acidic environment, respectively.

Boppana et al. [59] have synthesized the pH-sensitive polyacrylamide-grafted-tamarind seed polysaccharide (PAM-g-TSP) copolymer under microwave irradiation in the presence of CAN as an initiator. The IPN microbeads were also synthesized from hydrolyzed PAM-g-TSP copolymer and sodium alginate (NaAlg) through a dual cross-linking process for controlled delivery of ketoprofen avoiding its release in stomach environment. The IPN microbeads were tested *in vitro* and *in vivo* and showed pH-dependent swelling/shrinking with the changing pH of the medium. The synthesized products were characterized with the help of FTIR, ^{1}H-NMR, TGA, DSC, elemental analysis, and SEM techniques.

Pal and Pal [60] have synthesized an amphiphilic graft copolymer from tamarind gum and poly(methyl methacrylate), TG-g-PMMA, using atom transfer radical polymerization (ATRP) method in the presence of CuBr/bpy (copper bromide/Bipyrridine) catalyst at 65°C. The effects of various reaction conditions viz. concentrations of monomer and catalyst, time, and temperature on the graft copolymerization reaction were also investigated. FTIR and ^{1}H-NMR spectroscopic methods, DLS, TGA, and FESEM studies were carried out in order to investigate the structural and surface properties of the synthesized graft copolymer. The rheological study also was carried out to analyze the gel characteristic property of the copolymer. The synthesized copolymer exhibited an excellent sorption ability towards toxic MB and Congo red (CR) dyes. The adsorbed MB and CR dye on TG-g-PMMA surface was found to be desorbed efficiently and the copolymer was found to be reused completely.

Trivedi et al. [61] have studied an unreported photo-initiated graft copolymerization of methyl acrylate (MA) onto Na-PCMTKP ($\overline{DS} = 0.15$) using CAN as a photo-initiator. The influence of reaction variables on the grafting yields was studied and the optimal reaction conditions were evaluated. The efficiency of the photo-initiator was also studied by carrying out photo-graft copolymerization of MA onto Na-PCMTKP in the presence and absence of ultraviolet radiations. The evaluated optimal reaction conditions have also been utilized to study the effect of reaction medium on photo-grafting. It was observed that the reaction medium played an important role in photo-graft copolymerization reaction and the magnitude of grafting differed significantly with the type and ratio of solvent used in the mixture. The evidence of photo-grafting was ascertained by spectral (^{1}H-NMR and FTIR), SEM, and thermal (TGA/DSC) techniques. The synthesized graft copolymer, Na-PCMTKP-g-PMA, after treating with hydroxylamine in the alkaline medium may find its potential application as an adsorbent for metal ion sorption.

Choudhury et al. [62] have synthesized CMT-based graft copolymer with Hydroxy ethylmethacrylate (HEMA) (CMT-g-PHEMA) using BPO (benzoyl peroxide) as an initiator and prepared a set of hydrogels having HEMA and CMT in different mole ratios. These CMT:HEMA-based hydrogels were characterized by different physicochemical techniques viz. XRD, SEM, and dynamic light scattering (DLS). The hydrogel was found to be suitable for the growth of skin keratinocytes *in vitro* and in *ex-vivo* applications in whole animals. This study was undertaken with a view to using this hydrogel in the context of skin tissue engineering and skin regeneration.

Nandi et al. [63] have synthesized polyacrylamide grafted tamarind seed gum (PAM-g-TSG) employing a microwave-assisted method using CAN as a free radical initiator. The influence of the concentrations of monomer and CAN, as well as microwave irradiation time (MW) on grafting, was studied. The characterization of the graft copolymer was carried out using elemental analysis, FTIR, solid-state ^{13}C NMR, DSC, TGA, XRD, Viscosity, and SEM studies. The synthesized graft copolymer was evaluated for its flocculating property in paracetamol suspension. The non-toxic and biodegradable natures of the graft copolymer were also studied.

Trivedi et al. [64] have studied the functionalization of Na-PCMTKP ($\overline{DS}$ = 0.15) through photo-grafting with MMA and ethyl methacrylate (EMA) using CAN as a photo-initiator. The optimization of effective reaction conditions was established in order to achieve maximum grafting yields, and the influence of the various reaction conditions on the grafting yields was studied. The experimental results were analyzed in light of the proposed kinetic scheme and were found to be in good agreement. The values of the overall activation energy obtained in the case of photo-grafting of MMA and EMA were found to be 8.10 kJ/mol and

10.49 kJ/mol, respectively, indicating the ease of the occurrence of photo-grafting reaction in the case of MMA compared to EMA. The reactivity of both the monomers (MMA and EMA) towards photo-grafting was also compared on the basis of their structures as well as evaluated R_p (overall rate of polymerization) values at various monomers (MMA or EMA) and photo-initiator concentrations. A plausible explanation for the observed order of reactivity was provided. The FTIR, TGA, and SEM techniques were used to provide proof of grafting. The synthesized graft copolymers (Na-PCMTKP-g-PMMA and Na-PCMTKP-g-PEMA) may be used for environmental and pharmaceutical applications.

Trivedi et al. [65] have established the optimum reaction conditions in the case of UV-radiation-induced grafting of AN onto Na-PCMTKP ($\overline{DS}$ = 0.15) using CAN as a photo-initiator. The maximum values of %G and %GE were found to be 361.16% and 89.33%, respectively. A suitable mechanism of UV-radiation-induced grafting was also proposed. The graft copolymer was characterized by FTIR, TGA, DSC, and SEM techniques. The DSC results of the graft copolymer sample were analyzed in terms of Ozawa and Kissinger's methods for evaluating various kinetic parameters. The values of the rate constant and the half-life obtained by the Ozawa and the Kissinger methods were found to be reasonably in good agreement with each other.

CONFLICT OF INTEREST

The authors declare no conflict of interest.

KEYWORDS

- **carboxymethyl tamarind**
- **grafting**
- **sodium salt of partially carboxymethylated tamarind kernel powder (Na-PCMTKP)**
- **superabsorbent hydrogel (H-Na-PCMTKP-g-PAN)**
- **tamarind kernel powder (TKP)**
- **tamarind seed powder (TSP)**

REFERENCES

1. Bagul, M., Sonawane, S. K., & Arya, S. S., (2015). Tamarind seeds: Chemistry, technology, application and health benefits: A review. *Indian Food Industry Mag., 34*, 28–35.
2. Rao, Y. S., & Mathew, K. M., (2012). Tamarind (*Tamarindus indica L.) res*earch: A review. *Handbook of Herbs and Species* (pp. 512–533). Woodhead Publishing Limited.
3. Ferrara, L., (2019). Nutritional and pharmacological properties of *Tamarindus indica L. J. Nutr. Food Sci. Forecast. 2*, 1–5.
4. Coronel, R. E., (1991). *Tamarindus indica L. In:* Verheij, E. W. M., & Coronel, R. E., (eds.), *Plant Resources of Southeast Asia* (pp. 298–301). Wageningen, Pudoc. No. 2. Edible fruits and nuts. PROSEA Foundation, Bogor, Indonesia.
5. Shankaracharya, N. B., (1998). Tamarind–chemistry, technology and uses: A critical appraisal. *J. Food Sci. Technol., 35*, 193–208.
6. Khoja, A. K., & Halbe, A. V., (2001). *Scope for the use of Tamarind Kernel Powder as a thickener in textile printing* (Vol. 44, pp. 403–407). Man-Made Textiles in India.
7. Razavi, M., Nyamathulla, S., Karimian, H., Moghadamtousi, S. Z., & Noordin, M. I., (2014). Hydrogel polysaccharides of tamarind and xanthan to formulate hydrodynamically balanced matrix tablets of famotidine. *Molecules, 19*, 13909–13931.
8. Siddhuraju, P., Vijayakumari, K., & Janardhanan, K., (1995). Nutritional and antinutritional properties of the underexploited legumes *Cassia laevigata* Willd. and *Tamarindus Indica L. J. Food Compos. Anal., 8*, 351–362.
9. Patil, S. J., & Nadagouder, B. S., (1997). Industrial products from *Tamarindus Indica. Proceedings of National Symposium on Tamarindus indica L.* (pp. 151–155).Tirupati (A.P.), organized by the forest department of Andhra Pradesh.
10. Gidley, M. J., Lillford, P. J., Rowlands, D. W., Lang, P., Dentini, M., Crescenzi, V., Edwards, M., et al., (1991). Structure and solution properties of tamarind-seed polysaccharide. *Carbohydr. Res., 214*, 299–314.
11. Rao, P. S., & Srivastava, H. C., (1973). Tamarind. In: Whistler, R. L., (ed.), *Industrial Gums* (2nd edn., pp. 369–411). Academic Press, New York.
12. Kooiman, P., (*1961*). The constitution of Tamarindus-amyloid. *Rec. Trav. Chim. Pays Bas., 80, 849–865.*
13. Kumar, C. S., & Bhattacharya, S., (2008). Tamarind seed: Properties, processing and utilization. *Crit. Rev. Food Sci., 48,* 1–20.
14. Mishra, A., & Malhotra, A. V., (2009). Tamarind xyloglucan: A polysaccharide with versatile application potential. *J. Mater. Chem., 19*, 8528–8536.
15. Joseph, J., Kanchalochana, S. N., Rajalakshmi, G., Hari, V., & Durai, R. D., (2012). Tamarind seed polysaccharide: A promising natural excipient for pharmaceuticals. *Int. J. Green Pharm., 6,* 270–278.
16. Manchanda, R., Arora, S. C., & Manchanda, R. V., (2014). Tamarind seed polysaccharide and its modifications-versatile pharmaceutical excipients: A review. *Int. J. Pharm. Tech. Res., 6*, 412–420.
17. Sahoo, R., & Sahoo, S., (2015). Natural polymer of tamarind seed: A prospective carrier for ocular drug delivery. *Eur. Sci. J., 11*, 267–275.

18. Shukla, A. K., Bishnoi, R. S., Kumar, M., Fenin, V., & Jain, C. P., (2018). Applications of tamarind seeds polysaccharide-based copolymers in controlled drug delivery: An overview. *Asian J. Pharm. Pharmacol., 4*, 23–30.
19. Kuru, P., (2014). *Tamarindus indica and its* health related effects. *Asian Pac. J. Trop. Biomed., 4,* 676–681.
20. Rao, P. S., & Beri, R. M., (1955). Acetylation of tamarind seed jellose. *Proc. Indi. Acad. Sci., 42A*, 199–203.
21. Kaur, H., Yadav, S., Ahuja, M., & Dilbaghi, N., (2012). Synthesis, characterization and evaluation of thiolated tamarind seed polysaccharide as a mucoadhesive polymer. *Carbohydr. Polym., 90*, 1543–1549.
22. Mahajan, H. S., Tyagi, V. K., Patil, R. R., & Dusunge, S. B., (2013). Thiolated xyloglucan: Synthesis, characterization and evaluation as mucoadhesive in situ gelling agent. *Carbohydr. Polym., 91*, 618–625.
23. Lang, P., Masci, G., Dentini, M., Crescenzi, V., Cooke, D., Gidley, M. J., Fanutti, C., & Reid, J. S. G., (1992). Tamarind seed polysaccharide: Preparation, characterization and solution properties of carboxylated, sulphated and alkylaminated derivatives. *Carbohydr. Polym., 17*, 185–198.
24. Leppänen, Ann-Sofie, Xu, C., Parikka, K., Eklund, P., Sjöholm, R., Brumer, H., Tenkanen, M., & Willför, S., (2014). Targeted allylation and propargylation of galactose-containing polysaccharides in water. *Carbohydr. Polym., 100*, 46–54.
25. Goyal, P., Kumar, V., & Sharma, P., (2008). Cyanoethylation of tamarind kernel powder. *Starch/Starke, 60*, 41–47.
26. De Freitas, R. A., Busato, A. P., Mitchell, D. A., & Silveira, J. L. M., (2011). Degalatosylation of xyloglucan: Effect on aggregation and conformation, as determined by time dependent static light scattering, HPSEC–MALLS. *Carbohydr. Polym., 83*, 1636–1642.
27. Sumathi, S., & Ray, A. R., (2002). Release behavior of drugs from tamarind seed polysaccharide tablets. *J. Pharm. Pharm. Sci., 5*, 12–18.
28. Babu, P. R., Mathai, M. T., Julius, A., & Madhu, V., (2019). Dual property of tamarind seed polysaccharide aid wound healing. *International Journal of Advanced Science and Technology, 28,* 1130–1141.
29. Shaw, G. S., Uvanesh, K., Gautham, S. N., Singh, V., Pramanik, K., Banerjee, I., Kumar, N., & Pal, K., (2015). Development and characterization of gelatine-tamarind gum/carboxymethyl tamarind gum-based phase-separated hydrogels: A comparative study. *Des. Monomers Polym., 18*, 434–450.
30. Ajovalasit, A., Sabatino, M. A., Todaro, S., Alessi, S., Giacomazza, D., Picone, P., Carlo, M. D., & Dispenza, C., (2018). Xyloglucan based hydrogel films for wound dressing: Structure-property relationships. *Carbohydr. Polym., 179*, 262–272.
31. Prabhanjan, H., (1989). Studies on modified tamarind kernel powder. Part I: Preparation and physicochemical properties of sodium salt of carboxymethyl derivatives. *Starch/Starke, 41*, 409–414.
32. Goyal, P., Kumar, V., & Sharma, P., (2007). Carboxymethylation of tamarind kernel powder. *Carbohydr. Polym., 69*, 251–255.
33. Pal, S., Sen, G., Mishra, S., Dey, R. K., & Jha, U., (2008). Carboxymethyl tamarind: Synthesis, characterization and its application as novel drug-delivery agent. *J. Appl. Polym. Sci., 110*, 392–400.

34. Ponnikornkit, B., Ngamsalak, C., Huanbutta, K., & Sittikijyothin, W., (2015). Swelling behavior of carboxymethylated tamarind gum. *Adv. Mat. Res., 1060*, 137–140.
35. Meenakshi, & Ahuja, M., (2015). Metronidazole loaded carboxymethyl tamarind kernel polysaccharide-polyvinyl alcohol cryogels: Preparation and characterization. *Int. J. Biol. Macromol., 72*, 931–938.
36. Bhattacharya, A., Rawlins, J. W., & Ray, P., (2009). *Polymer Grafting and Crosslinking*. Wiley, New Jersey.
37. Hebeish, A., & Guthrie, J. T., (1981). *The Chemistry and Technology of Cellulosic Copolymers* (p. 351). Springer-Verlag, Heidelberg, New York.
38. Omidian, H., & Park, K., (2012). In: Juergen, S., Ronald, S., & Michael, R., (eds.), *Fundamentals and Applications of Controlled Release Drug Delivery* (pp. 75–106). Springer, New York, NY.
39. Bhatia, J. K., Kaith, B. S., & Kali, S., (2013). Polysaccharide hydrogels: Synthesis, characterization and applications. Kalia, S., & Sabaa, M. W., (eds.), *Polysaccharide Based Graft Copolymers* (pp. 271–290). Springer-Verlag Berlin Heidelberg, Ch.7.
40. Guilherme, M. R., Aouada, F. A., Fajardo, A. R., Martins, A. F., Paulino, A. T., Davi, M. F. T., Rubira, A. F., & Muniz, E. C., (2015). Superabsorbent hydrogels based on polysaccharides for application in agriculture as soil conditioner and nutrient carrier: A review. *Eur. Polym. J., 72*, 365–385.
41. Mishra, A., & Bajpai, M., (2006). Graft copolymerization of polyacrylamide onto tamarind mucilage. *J. Macromol. Sci., 43A*, 315–326.
42. Mishra, A., Bajpai, M., Pal, S., Agrawal, M., & Pandey, S., (2006). *Tamarindus indica mucilage* and its acrylamide-grafted copolymer as flocculants for removal of dyes. *Colloid Polym. Sci., 285*, 161–168.
43. Goyal, P., Kumar, V., & Sharma, P., (2008). Graft copolymerization of acrylamide onto tamarind kernel powder in the presence of ceric ion. *J. Appl. Polym. Sci., 108*, 3696–3701.
44. Mishra, A., Clark, J. H., Vij, A., & Daswal, S., (2008). Synthesis of graft copolymers of xyloglucan and acrylonitrile. *Polym. Adv. Technol., 19*, 99–104.
45. Singh, V., Tripathi, D. N., Malviya, T., & Sanghi, R., (2009). Persulfate/ascorbic acid initiated synthesis of poly(acrylonitrile)-grafted tamarind seed gum: A potential commercial gum. *J. Appl. Polym. Sci., 111*, 539–544.
46. Goyal, P., Kumar, V., & Sharma, P., (2009). Graft copolymerization onto tamarind kernel powder: Ceric (IV)–initiated graft copolymerization of acrylonitrile. *J. Appl. Polym. Sci., 114*, 377–386.
47. Sen, G., & Pal, S., (2009). Polyacrylamide grafted carboxymethyl tamarind (CMT-g-PAM): Development and application of a novel polymeric flocculant. *Macromol. Symp., 277*, 100–111.
48. Ghosh, S., Sen, G., Jha, U., & Pal, S., (2010). Novel biodegradable polymeric flocculant based on polyacrylamide-grafted tamarind kernel polysaccharide. *Bioresour. Technol., 101*, 9638–9644.
49. Ghosh, S., Jha, U., & Pal, S., (2011). High performance polymeric flocculant based on hydrolyzed polyacrylamide grafted tamarind kernel polysaccharide (Hyd. TKP-g-PAM). *Bioresour. Technol., 102*, 2137–2139.

50. Shailaja, T. S., Latha, K., Yarragudi, S., & Alkabab, A. M., (2012). A novel bioadhesive polymer: Grafting of tamarind seed polysaccharide and evaluation of its use in buccal delivery of metoprolol succinate. *Der Pharmacia Lettre., 4*, 487–508.
51. Mishra, A., & Malhotra, A. V., (2012). Graft copolymers of xyloglucan and methyl methacrylate. *Carbohydr. Polym., 87*, 1899–1904.
52. Ghosh, S., & Pal, S., (2013). *Modified tamarind kernel polysaccharide: A novel matrix for control release of aspirin. Int. J. Biol. Macromol., 58*, 296–300.
53. Trivedi, J. H., Jivani, J. R., Patel, K. H., & Trivedi, H. C., (2013). Modification of sodium salt of partially carboxymethylated tamarind kernel powder through grafting with acrylonitrile: Synthesis, characterization and swelling behavior. *Chinese J. Polym. Sci., 31*, 1670–1684.
54. Ahuja, M., Kumar, S., & Kumar, A., (2013). Tamarind seed polysaccharide-g-poly(N-vinyl-2-pyrrolidone): Microwave-assisted synthesis, characterization, and evaluation as mucoadhesive polymer. *Int. J. Polym. Mater., 62*, 544–549.
55. Trivedi, J. H., (2013). Synthesis, characterization and swelling behavior of superabsorbent hydrogel from sodium salt of partially carboxymethylated tamarind kernel powder-g-PAN. *J. Appl. Polym. Sci., 129*, 1992–2003.
56. Meenakshi, Ahuja, M., & Verma, P., (2014). MW-assisted synthesis of carboxymethyl tamarind kernel polysaccharide-g-polyacrylonitrile: Optimization and characterization. *Carbohydr. Polym., 113*, 532–538.
57. Del Real, A., Wallander, D., Maciel, A., Cedillo, G., & Loza, H., (2015). Graft copolymerization of ethyl acrylate onto tamarind kernel powder, and evaluation of its biodegradability. *Carbohydr. Polym., 117*, 11–18.
58. Pal, A., & Pal, S., (2016). Synthesis of copolymer derived from tamarind kernel polysaccharide (TKP) and poly(methyacrylic acid) via SI-ATRP with enhanced pH triggered dye removal. *RSC Adv., 6*, 2958–2965.
59. Boppana, R., Kulkarni, R. V., Mohan, G. K., Mutalik, S., & Aminabhavi, T. M., (2016). In vitro and in vivo assessment of novel pH-sensitive interpenetrating polymer networks of a graft copolymer for gastro-protective delivery of ketoprofen. *RSC Adv., 6*, 64344–64356.
60. Pal, A., & Pal, S., (2017). Amphiphilic copolymer derived from tamarind gum and *poly* (methyl methacrylate) via ATRP towards selective removal of toxic dyes. *Carbohdyr. Polym., 160*, 1–8.
61. Trivedi, J. H., Joshi, H. A., & Trivedi, H., (2017). Na-PCMTKP-g-PMA: Photo-induced synthesis and characterization. *International Journal of Research in Engineering and Applied Sciences, 7*, 39–55.
62. Choudhury, P., Kumar, S., Singh, A., Kumar, A., Kaur, N., Sanyasi, S., Chawla, S., et al., (2018). Hydroxyethyl methacrylate grafted carboxy methyl tamarind (CMT-g-HEMA) polysaccharide based matrix as a suitable scaffold for skin tissue engineering. *Carbohydr. Polym., 189*, 87–98.
63. Nandi, G., Changder, A., & Ghosh, L. K., (2019). Graft-copolymer of poly-acrylamide-tamarind seed gum: Synthesis, characterization and evaluation of flocculating potential in peroral paracetamol suspension. *Carbohydr. Polym., 215*, 213–225.

64. Trivedi, J. H., Joshi, H. A., & Trivedi, H. C., (2021). Synthesis and characterization of photo-graft copolymers of sodium salt of partially carboxymethylated tamarind kernel powder. *Macromol. Symp., 398*.
65. Trivedi, J. H., Joshi, H. A., & Trivedi, H. C., (2021). Graft copolymerization of acrylonitrile onto sodium salt of partially carboxymethylated tamarind kernel powder using ceric ammonium nitrate as a photo-initiator in an aqueous medium under ultraviolet–radiation. *Macromol. Symp., 398*, 1–13.
66. Trivedi, J.H. (2013), Synthesis, characterization, and swelling behavior of superabsorbent hydrogel from sodium salt of partially carboxymethylated tamarind kernel powder-g-PAN. J. Appl. Polym. Sci., 129: 1992-2003. https://doi.org/10.1002/app.38910

CHAPTER 5

Electrical Properties of $NA_2 PB_2 LA_2 W_2 TI_4 V_4O_{30}$ Ferroelectric Ceramic

PIYUSH R. DAS,[1] S. BEHERA,[2] S. K. MOHANTY,[3] and KHUSBOO AGRAWAL[1]

[1]*Department of Physics, Veer Surendra Sai University of Technology, Burla–768018, Sambalpur, Odisha, India, E-mail: prdas63@gmail.com (P. R. Das)*

[2]*Department of Physics, School of Applied Sciences, Centurion University of Technology and Management, Bhubaneswar–752050, Odisha, India*

[3]*Department of Physics, M. H. D. College, Chhatia, Jajpur, Odisha, India*

ABSTRACT

Polycrystalline ceramic $Na_2Pb_2La_2W_2Ti_4V_4O_{30}$ is synthesized using a mixed oxide route. The preparation condition was optimized using the thermogravimetry (TG) technique. The room temperature XRD study reveals the orthorhombic crystal structure. SEM shows uniform distribution of the grains. Variation of ε_r and tanδ with temperature shows the possible existence of ferroelectricity. Complex Impedance Spectroscopic study reveals a strong connection linking its microstructure and electrical parameters. AC conductivity curve reveals that Jonscher's power law is obeyed. The hysteresis loop confirms ferroelectricity in the investigated sample.

Advances in Diverse Applications of Polymer Composites: Synthesis, Application, and Characterization.
Suji Mary Zachariah, Yang Weimin, Maciej Jaroszewski, & Sabu Thomas (Eds.)

5.1 INTRODUCTION

Ferroelectric ceramics have been intensively investigated due to their functional properties, including switchable polarization, piezoelectricity, pyroelectricity, high nonlinear optical behavior, and high nonlinear dielectric behavior [1–5] which are applicable in electronic devices, namely capacitors, transducer, actuators, etc. Tungsten-bronze crystals with $(A_1)_2(A_2)_4(C)_4(B_1)_2(B_2)_8O_{30}$ formula are the 2nd largest class of ferroelectrics. Its electrical properties and potential application have resulted in remarkable attention for research [6]. The tetragonal unit cell consists of distorted BO_6 octahedra linked by their corners in a manner so as to form three different types of tunnels running right through the structure. The physical features of materials are changed considerably by the addition of different ions at A and B positions [7]. It has also been evident from the literature that ferroelectricity in a similar compound with niobate, i.e., $Na_2Pb_2La_2W_2Ti_4Nb_4O_{30}$ [9] and tantalates, i.e., $Na_2Pb_2La_2W_2Ti_4Ta_4O_{30}$ [10] has already been reported. Since the calcination temperature of niobates and tantalates is high compared to that of vanadates, an attempt has been made to prepare a complex vanadate $Na_2Pb_2La_2W_2Ti_4V_4O_{30}$. Further, we have reported studies on dielectric permittivity, conductivity, and impedance as a function of temperature and frequency have been carried out.

5.2 EXPERIMENTAL DETAILS

Polycrystalline sample $Na_2Pb_2La_2W_2Ti_4V_4O_{30}$ (hereafter abbreviated as NPLWTV) was synthesized by the solid-state reaction method. Highly pure compounds of Na_2CO_3, PbO, La_2O_3, TiO_2, V_2O_5, and WO_3 (99.9%, M/s Loba Chemie Pvt. Ltd., India) were taken in a stoichiometric amount which was then mixed and grounded. About 5 mg of the physically mixed sample was taken for thermal analysis (PERKIN ELMER) at a temperature ranging from 50–850°C. Then calcination of the mixture was carried out for 6 hours at 650°C. XRD pattern of calcined powder was obtained from an X-ray powder diffractometer (Rigaku Miniflex) with CuK_α radiation ($\lambda = 1.5405Å$) at a Bragg angle ($20° \leq 2\theta \leq 80°$). The fine powder was pelletized using a hydrolytic press, and sintering was done at 675°C for 4 hours. Surface morphology was realized using JEOL JSM-580 scanning electron microscope (SEM). Pellets were electrode with highly pure air-drying silver paste and heated at 150°C for 1.5 hours. The effect of frequency (1 kHz–1 MHz) on electrical properties (dielectric, impedance, etc.), of the sintered pellet was measured employing computer-controlled HIOKI 3532 LCR in a set of temperatures. The hysteresis loop at room temperature of the pooled sample was realized by a P-loop tracer (M/S. Radiant Technologies Inc., NM, USA).

5.3 RESULTS AND DISCUSSION

5.3.1 THERMAL ANALYSIS

Differential thermogravimetry analysis (DTA) and thermogravimetry analysis (TGA) is represented in Figure 5.1(b). It is seen from the TGA pattern about 2% of the mass is lost at around 250°C due to the disappearance of surface-absorbed moisture. In the 2nd stage, mass loss (~6%) in the 250°C–400°C temperature range is owed to the decomposition of carbonate with the evolution of CO_2 [8]. In the 3rd stage, small mass loss at a temperature range of 400°C–550°C occurs due to the evolution of a small amount (if any) of CO_2 gas. Above 550°C, no marked loss of mass was observed, implying that the reaction is complete and a new compound is formed. It is found that the TGA observations have close conformity with that of DTG with the appearance of peaks for each step of reaction [8, 11]. Thus, the material was finally calcined at 650°C.

5.3.2 STRUCTURAL/MICROSTRUCTURE

The room temperature XRD pattern is shown in Figure 5.1(a). The diffraction pattern obtained has incisive peaks, which are distinct from the components bearing improved uniformity and crystallinity, thereby confirming the formation of a single-phase compound [12]. The entire study of obtained XRD peak was carried out through programmed computer software "POWD" [13]. Based on a fair agreement between observed (obs.) and calculated (cal.) inter-planner spacing (d), an orthorhombic crystal system was selected. The lattice parameters (a, b, c) and volume (V) obtained are 19.5094(07)Å, 19.0978(07)Å, 3.7609(04), and 1401.26 Å^3, respectively.

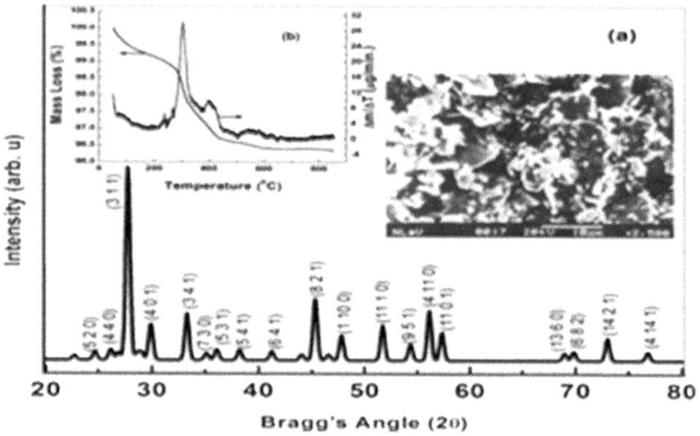

FIGURE 5.1 (a) Room temperature XRD; (b) TGA and SEM of NPLWTV.

Additionally, crystalline or particle size (P) was derived using Scherer's formula [12]:

$$P_{\mathrm{hkl}} = \frac{k\lambda}{\beta_{\frac{1}{2}} \cos\theta_{hkl}} \quad (1)$$

The average particle size P_{hkl} is obtained to be approximately 31 nm.

Figure 5.1 (inset) represents the surface morphology of NPLWTV. It is seen that grains are elongated and systematically disseminated over the complete surface. Grain size ranges from 2–5 μm.

5.3.3 DIELECTRIC PROPERTIES

Response of relative dielectric constant ε_r with temperature is represented in Figure 5.2. ε_r value rises with an increase in temperature and reaches the highest value at 393 K (called phase transition temperature T_c), later drops till 452 K and then rises. The rise in ε_r below Tc occurs due to electron-phonon interaction, dipolar ordering, etc. [14]. Beyond T_c decrease in ε_r can be attributed to paraelectric behavior. The rise of ε_r beyond 452 K may be owed to the existence of space charge polarization which is thermally activated. At lower frequency (i.e., 1 kHz), values of dielectric constant largely own to charge aggregation at grain boundaries [15]. An increase in frequency with a decrease in dielectric constant is owed to deduction in space charge polarization. The incisiveness of curves falls with an increase in frequency owns to the occurrence of dispersed phase transition.

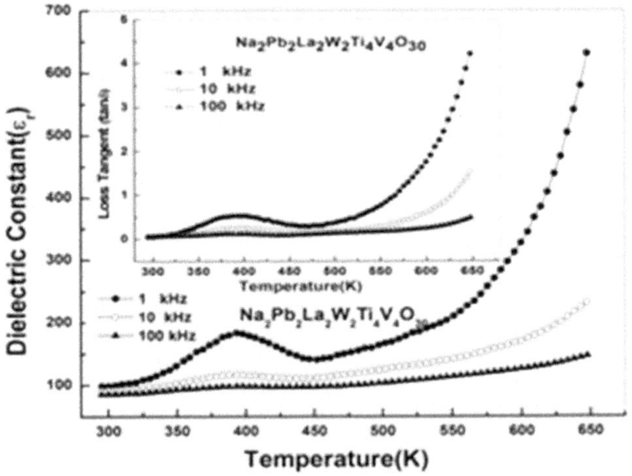

FIGURE 5.2 ε_r and tan δ vs. temperature at different frequencies.

The effect of temperature on dielectric loss (tan δ) is represented in Figure 5.4 (inset). Response of tanδ with temperature is analogous to that of dielectric constant with temperature, but the rise in tanδ at low temperature is found to be small, and at high temperature, it rises significantly. The rise in tan δ at higher temperature can be owed to the diffusion of charge carriers which is thermally activated, and some defects in the sample [16].

5.3.4 *HYSTERESIS*

Room temperature hysteresis confirming ferroelectricity is represented in Figure 5.3 with 2Pr = 0.130 μCm^{-2} and Ec = 7.58 KV/cm.

5.3.5 *IMPEDANCE ANALYSIS*

Complex impedance spectrography is a distinctive, robust, and non-destructive method for identifying electrical responses in a broad series of temperature and frequency. In order to measure output, an AC signal is applied through the pellet. Figure 5.4 represents the change in Z'' with frequency. The value of Z'' rises to reach the maximum value and then falls rapidly with an increase in temperature, thereby explaining the occurrence of the relaxation process [17]. Further, the existence of the relaxation phenomenon is established by an increment in the widening of the curve with an increase in temperature. At higher temperatures relaxation process originates owing to the occurrence of voids and imperfection, whereas at low temperatures, it is due to immobile species [18, 19].

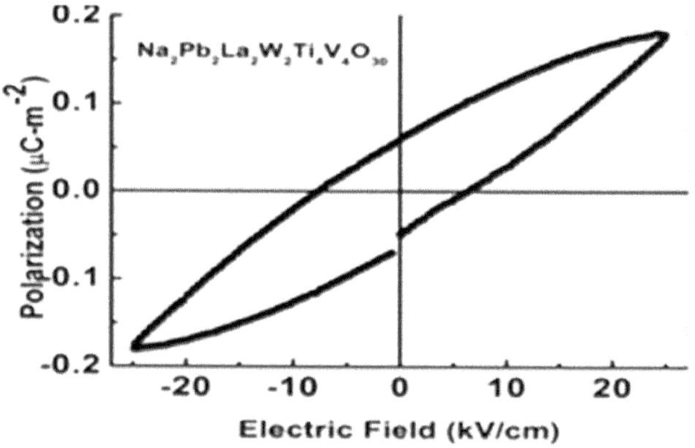

FIGURE 5.3 Hysteresis loop.

Nyquist plot (Z' vs Z'') is represented in Figure 5.4 (inset). A single semicircular curve has been obtained, thereby representing the contribution of the bulk effect. Semicircular arc at low frequency is owned to grain boundary effect, whereas the bulk effect is seen at the high-frequency semicircular arc. With the increase in temperature, semicircular intercept displaces about a lower value of Z', suggesting a decrease in bulk resistance and the grain boundary. Such electric behavior can be described considering an identical circuit consisting of parallel connection of RC circuits. Further, depressed arcs with centers below the real axis reveal non-Debye type behavior [20].

5.3.6 AC CONDUCTIVITY

The influence of frequency on AC conductivity is presented in Figure 5.5. The material obeys the Jonscher universal power law [19], and the conductivity is found to be increasing with increment in temperature and frequency. Dispersion is found to be more in low temperatures than in high temperatures, which may be attributed to electrode polarization. Again, the mobility of charge carriers decreases due to less concentration of oxygen vacancies, thereby reducing conductivity at low frequencies.

At high frequencies, the conductivity curves tend to merge, and thus dispersion becomes temperature and frequency independent.

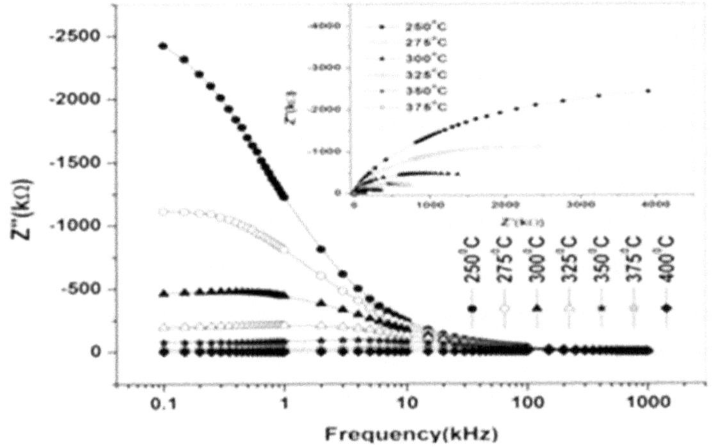

FIGURE 5.4 Z" vs. frequency and Nyquist ($Z' - Z''$) plot at different temperatures.

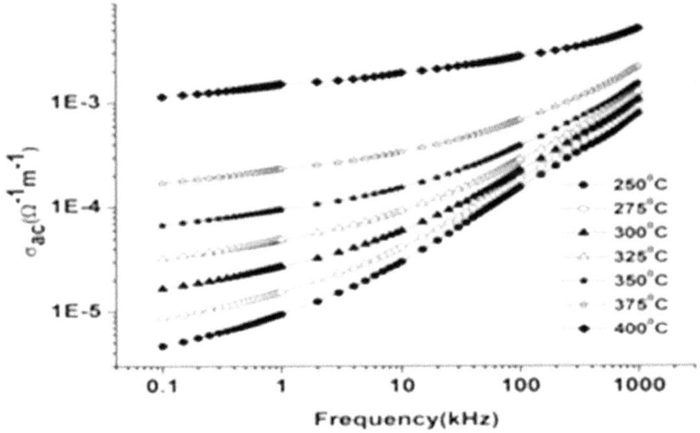

FIGURE 5.5 Variation of AC conductivity σ_{AC} with frequency at different temperatures.

5.4 CONCLUSION

The structural and dielectric property of NPLWTV having an orthorhombic system has been explored over a broad range of frequencies and temperatures. The hysteresis loop confirms ferroelectricity in the system.

SEM studies show that surfaces have uniform grain distribution. Complex impedance plot represents that electrical transport phenomenon is due to temperature-dependent relaxation phenomenon and bulk effect, i.e., NTCR (negative temperature coefficient of resistivity) type-response. AC conductivity proves that Jonscher's power law is satisfied.

KEYWORDS

- **conductivity**
- **ferroelectric**
- **hysteresis**
- **impedance**
- **scanning electron microscope**
- **thermogravimetry analysis**
- **X-ray diffraction**

REFERENCES

1. Bain, A. K., & Chand, P., (2017). *Ferroelectricity: Principles and Applications*. Wiley.
2. Lei, C., et al., (2020). *Cer. Int., 46*, 6108.
3. Suchanicz, J., et al., (2018). *Ferroelectrics, 524,* 9.
4. Jiang, W., et al., (2005). *J. Appl. Phys., 97*, 094106/1.
5. Gao, T. T., et al., (2016). *Mater. Chem. Phys., 181*, 47.
6. Shudong, X., et al., (2020). *Cer. Int., 46*, 13997.
7. Neurgaonkar, R. R., et al., (1990). *Mater. Res. Bull., 25*, 959.
8. Das, P. R., et al., (2013). *Phase Transitions, 86*, 1267.
9. Biswal, L., et al., (2012). *J. Electroceram, 29*, 204.
10. Das, P. R., et al., (2012). *J. Adv. Cer., 1*(3), 232.
11. Kolta, G. A., et al., (1973). *Thermochim. Acta, 6*, 165.
12. Klug, H. P., & Alexander, L. E., (1974). *X-ray Diffraction Procedures* (p. 966). Wiley, Chichester, UK.
13. Wu, E. *POWD, An interactive Powder Diffraction Data Interpretation and Indexing Program,* Version 2.1, School of Physical Science, Finders University of South Australia, Bedford Park, S.A.5042, Australia.
14. Kityk, I. V., et al., (2001). *J. Appl. Phys., 90*, 5542.
15. Jawahar, K., et al., (2008). *Mater. Lett., 62*, 911.

16. Van, U. L. G., et al., (1969). *Mater. Res. Bull., 463.*
17. Pradhani, N., et al., (2019). *Mater. Res. Bull., 119*, 110566.
18. Suman, C. K., et al., (2006). *J. Mater. Sc., 41*, 369.
19. Jonscher, A. K., (1977). *Nature, 267*, 673.
20. Deng, G., et al., (2005). *App. Phy. Let., 87*, 3.

CHAPTER 6

Investigation of Dielectric and Ferroelectric Properties of PVDF/0.5Ba $(Zr_{0.2}Ti_{0.8})O_3$–0.5$(Ba_{0.7}Ca_{0.3})TiO_3$ Composite

SAKTI PRASANNA MUDULI,[1] S. PARIDA,[1] S. K. ROUT,[2] and S. K. MAHAPATRA[3]

[1]*Department of Physics, C.V. Raman Global University, Bhubaneswar–752054, Odisha, India, E-mail: sabyasachiparida023@gmail.com (S. Parida)*

[2]*Department of Physics, Birla Institute of Technology, Mesra, Ranchi–835215, Jharkhand, India*

[3]*Center for Physical Sciences, Central University of Punjab, Bathinda–151001, Punjab, India*

ABSTRACT

(1-x)PVDF-(x)[0.5Ba$(Zr_{0.2}Ti_{0.8})O_3$–0.5$(Ba_{0.7}Ca_{0.3})TiO_3$] composite films with x = 0.1, 0.2, 0.3, 0.4, 0.5 were synthesized and the comparative dielectric properties and ferroelectric properties were studied. 0.5Ba$(Zr_{0.2}Ti_{0.8})O_3$–0.5$(Ba_{0.7}Ca_{0.3})TiO_3$ (BZT-BCT) ceramic was synthesized by conventional solid-state reaction method. After phase confirmation of calcined powder, it was added to poly(vinylidene fluoride) (PVDF) solution with different weight percentages, and composite films were prepared by solvent casting followed by hot pressing method.

Advances in Diverse Applications of Polymer Composites: Synthesis, Application, and Characterization. Suji Mary Zachariah, Yang Weimin, Maciej Jaroszewski, & Sabu Thomas (Eds.)

XRD, as well as FTIR analysis, shows the formation of the electroactive β-phase of PVDF, and all the samples have a fraction of τηε β-phase of more than 65%. The dielectric constant of polymer ceramic composite film increases from 10.2 to 19 for x = 0 to 0.5. All the polymer ceramic composite samples have a dissipation factor of less than 0.1. Ferroelectric hysteresis study shows that remnant polarization increases with an increase in ceramic content. The efficiency of the storage energy density changes from 90.12% to 64.38% whereas the recoverable storage energy density increases from 2.1 mJ/cm^3 for x = 0 to 5.51 mJ/cm^3 for x = 0.5.

6.1 INTRODUCTION

Recently polymer ceramic composite has drawn the attention of many researchers due to better electro-mechanical properties of composites than that of pure ceramics or polymers alone. Polymer ceramic composite receives excellent dielectric, ferroelectric, and piezoelectric properties from ceramic, along with flexibility, high breakdown strength, and good processing properties from polymer [1]. Polymer ceramic composites are widely used for sensors [2], capacitive storage devices [3], energy harvesting [4], and biomedical applications [5]. Since the discovery of the piezoelectricity of poly(vinylidene fluoride) (PVDF), PVDF and its copolymers have been used for composites for the last decades as they show overall good dielectric and ferroelectric properties [6]. PVDF shows at least four distinct polymorphs depending on the orientation of C-F and C-H bonds in the molecular chain. Among all these polymorphs, the β-phase has an all-trans (TTTT) planar zigzag structure [7]. It is the most electroactive phase as the dipole moments of C-F and C-H bonds add up to give an effective dipole moment in the direction perpendicular to the carbon backbone [6, 7].

$PbTiO_3$ [8], $Pb(Zr,Ti)O_3$ (PZT) [9, 10], etc., were preferred for composite preparation because of their good dielectric and piezoelectric properties. For example, Wankhade et al. [4] prepared PVDF-PZT composite by compression molding method and reported the results of energy harvest in different methods; Singh et al. [11] reported the dielectric and energy storage properties of PVDF-PZT composite prepared by solvent casting method. But due to the toxicity of lead oxide, now researchers go-ahead with lead-free ceramics that can replace lead-based ceramics for dielectric and piezoelectric applications [12, 13]. Many lead-free ceramic materials have been reported to act as an alternative to lead-based materials. However, these ceramic materials fall short of lead-based ceramics in terms of piezoelectric properties [12]. Recently it has been reported that the lead-free solid solution ceramics having morphotropic phase boundary (MPB) show piezoelectric properties comparable with the lead-based ceramics [14, 15]. The composition $[(1\text{-}x)Ba(Zr_{0.2}Ti_{0.8})O_3–x(Ba_{0.7}Ca_{0.3})TiO_3]$ (BZT-BCT) was found to be having piezoelectric property comparable to PZT and even having better dielectric and ferroelectric properties at its MPB ($x = 0.5$) [12, 13, 15]. It shows a high dielectric constant up to 3,900 (Room temperature) at 1 kHz and a piezoelectric strain coefficient of around 620 pC/N [12, 13]. Recently some works on PVDF-(BZT-BCT) have been reported. For example, Pandey et al. [16] reported the piezoelectric and temperature-dependent dielectric properties of PVDF-(BZT-BCT) prepared by the melt mixing method; Chi et al. [17] studied the microstructure and dielectric properties of solvent cast PVDF-(BZT-BCT) composite in which BZT-BCT was in nanofiber form.

As explained, some works have been reported on PVDF-(BZT-BCT) prepared by different methods (such as solvent casting, melt mixing, and compression molding), and a detailed discussion on ferroelectric properties of the composite prepared by solvent casting followed by the hot pressing method has not been reported so far. Hence this work represents the effect of BZT-BCT content in PVDF/(BZT-BCT) composite, prepared by solvent casting followed by hot pressing method, on dielectric and ferroelectric properties along with energy density.

6.2 SYNTHESIS AND CHARACTERIZATION METHODS

$0.5Ba(Zr_{0.2}Ti_{0.8})O_3–0.5(Ba_{0.7}Ca_{0.3})TiO_3$ ceramic was synthesized by solid-state reaction method using $BaZrO_3$ (Sigma Aldrich), $BaCO_3$ (Merck), $CaCO_3$ (Loba Chemie), TiO_2 (Loba Chemie) as raw materials. Phase confirmation was done after calcination at 1,350°C for 6 hours. PVDF-(BZT-BCT) composite was prepared by solvent casting followed by a hot pressing method. The calculated amount of PVDF(Alfa Aesar) was added to N,N-dimethylformamide (DMF) (CDH), and stirred for 30 minutes to get a transparent solution. BZT-BCT calcined powder of different weight percentages (0, 10, 20, 30, 40, and 50%) (Equivalent volume percentage given in Table 6.1) was added to that solution and ultra-sonicated till well dispersion of ceramic powder into the solution. The solutions were stirred at 50°C for 10 hours to get the viscous solution, then cast into a Petri dish and kept in a hot air oven at 100°C for 7 hours to get all the solvent evaporated. Then each solvent cast film was hot-pressed at 150°C and 75 $kgcm^{-2}$ pressure with the help of a lab-scale hydraulic hot press for 1 hour. Schematic diagram of the preparation procedure is given in Figure 6.1 and thicknesses of prepared samples are given in Table 6.1. Volume fraction equivalent of weight fraction is also given in Table 6.1. The volume fraction is calculated using the following Eqn. (1) [18].

$$V_f = \frac{W_f \cdot \rho_{PVDF}}{\rho_{BZTBCT} + W_f \left(\rho_{PVDF} - \rho_{BZTBCT}\right)} \times 100\% \tag{1}$$

where; V_f, W_f are volume fraction (percentage) and weight fraction, respectively; ρ_{PVDF} and ρ_{BZTBCT} are densities of PVDF (1.78 g/cm^3) and BZT-BCT (Experimental 5.51 g/cm^3), respectively. X-ray diffraction (XRD) measurements were performed using Bruker AXS D8 Advance with Cu-Kα (λ = 1.5406 A°). The samples were scanned in the 2θ range of 10° to 80°. Fourier transform infrared (FTIR) spectra were obtained via Thermo-scientific Nicolet iS 5 FT-IR spectrometer in the range 400 cm^{-1} to 1,300 cm^{-1}. The morphology of the samples was investigated using high-resolution scanning electron microscope (SEM; FEI-Quanta FEG 200F). Silver conducting paste was painted on both surfaces of each film for dielectric and ferroelectric measurements. Frequency-dependent dielectric data were obtained with the help of Solartron 1260A Impedance Analyzer in the frequency range from 10 Hz to 10^6 Hz. Ferroelectric hysteresis loops were obtained using radiant ferroelectric equipment (version: 5.7.0) at 10 Hz.

6.3 RESULT AND DISCUSSION

6.3.1 XRD

XRD patterns for BZT-BCT, PVDF film, and composite films are given in Figure 6.2(a)–(c), respectively. The phase formation of BZT-BCT was confirmed by XRD, followed by Rietveld refinement. The structural refinement was performed using the FullProf program. (FullProf-2000, version July 2001). Characteristic peaks of BZT-BCT are indexed and shown in Figure 6.2(a). The bifurcated peak at around 66° corresponds to both (220) and (202) and represents the coexistence of both rhombohedral and tetragonal phases [14]. Detailed crystallographic data is given in Table 6.2. No visible peak for the second phase is observed, and the pattern matches with the literature [14, 15], showing MPB type structure.

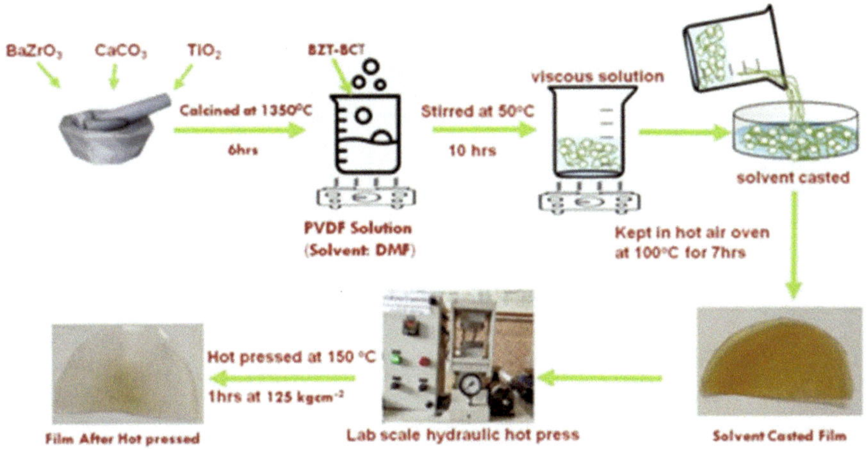

FIGURE 6.1 Schematic diagram of composite preparation.

TABLE 6.1 Equivalent Volume Percentages

Weight Percentage of BZT-BCT	Volume Percentage of BZT-BCT	Sample Name	Thickness (mm)	Fraction of β-Phase	Dielectric Constant at 1 kHz	Dissipation Factor at 1 kHz
0	0	PVDF	0.120	88.6%	10.2	0.042
10	3.47	PC10	0.480	69%	11.2	0.047
20	7.47	PC20	0.509	74.6%	12.6	0.074
30	12.16	PC30	0.526	65.5%	16.8	0.042
40	17.72	PC40	0.370	72.3%	17.7	0.041
50	24.42	PC50	0.335	79.6%	19	0.043

The XRD pattern of PVDF film shows a significant reduction in the A-phase peaks at 2θ values 17.7°, 18.5°, and 26.7° with characteristics (100), (020), (021), respectively in comparison to PVDF powder (inset of Figure 6.2(b)). The only clear peak at 20.4° confirms the formation of electroactive β-phase in the PVDF film [6, 7].

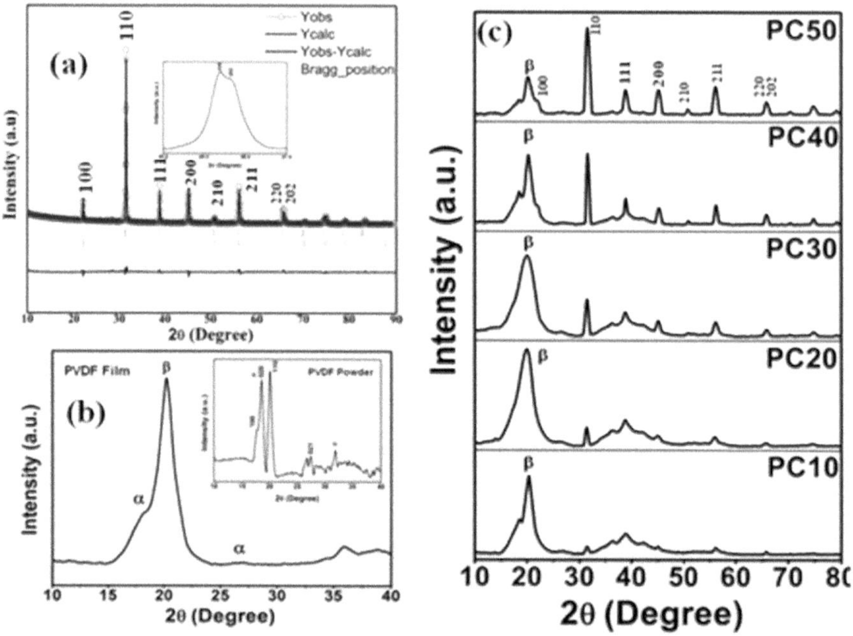

FIGURE 6.2 XRD patterns of (a) BZT-BCT; (b) PVDF film; (c) composites.

TABLE 6.2 Detailed Crystallographic Data of BZT-BCT Sample

	BZT-BCT	
a (Å)	5.703073	4.007139
b (Å)	5.703073	4.007139
c (Å)	7.004964	4.022909
Volume (Å)^3	197.312	64.597
Structure	Rhombohedral	Tetragonal
Space group	R3m	P4mm
	Rhombohedral = 25.60%	Tetragonal = 74.40%
χ^2	**2.59**	

All the composites show both PVDF phase and BZT-BCT phase. The peak at 2θ value 20.4°, which corresponds to β-phase of PVDF, is present in all the XRD patterns for composite films but with an increase of ceramic content, intensity of PVDF peak gradually reduces and the same for BZT-BCT increases. All the ceramic peaks are prominent in composites maintaining the β-phase of PVDF. Further, to verify the existence of β-phase and estimate its fraction in the composites, FTIR spectra were analyzed.

6.3.2 FTIR

FTIR spectra for BZT-BCT ceramic is represented in Figure 6.3(a). Absorbance sharp peak at wave number 430 cm^{-1} and broad peak in the range 500–700 cm^{-1} corresponds to Zr-O and Ti-O bonds, respectively [19]. Figure 6.3(b) and (c) show the FTIR spectra for PVDF film and PVDF powder, respectively, and it can be clearly noticed that characteristic α peaks of PVDF powder at 763 cm^{-1}, 795 cm^{-1}, 975 cm^{-1}, 1,065 cm^{-1}, 1,150 cm^{-1} and 1,210 cm^{-1} are diminished in PVDF film [6, 20]. The peaks at 840 cm^{-1}, 1,234 cm^{-1}, and 1,274 cm^{-1}, which correspond to electroactive phase of PVDF, are observed in PVDF film [6, 21]. During the processing of the PVDF powder to make the film, C-H, and C-F bonds reorient themselves by rotating around the carbon backbone. So, the polymorph changes from TGTG' to TTTT configuration, and this is the reason behind the diminishing of the intensity of phase peaks and rising of the peaks corresponding to β phase [7, 21, 20]. Figure 6.3(d) represents the FTIR spectra for PVDF-(BZT-BCT) composite with variation of ceramic content in PVDF matrix. Peaks of both ceramic (represented by 'c') and polymer (β-phase represented by β) are clearly identified from each spectrum and the intensity of characteristic absorbance peaks of PVDF gradually decreases with the increase of ceramic content in composite. The electroactive β-phase of PVDF affects the effective dielectric properties of the composite. Hence calculation of the fraction of β-phase in the samples is important for further study. The fraction of β-phase of PVDF for all the samples is calculated using Eqn. (2) and listed in Table 6.1.

$$F_{\beta} = \frac{I_{840}}{1.262 I_{763} + I_{840}} \times 100\% \qquad (2)$$

where; F_{β}, I_{840}, and I_{763} are fraction of β-phase, absorbance intensities of the peaks at wave number 840 cm^{-1} and 763 cm^{-1} [21, 22]. It can be observed from Table 6.1 that F_{β} of all the composites are more than 65%.

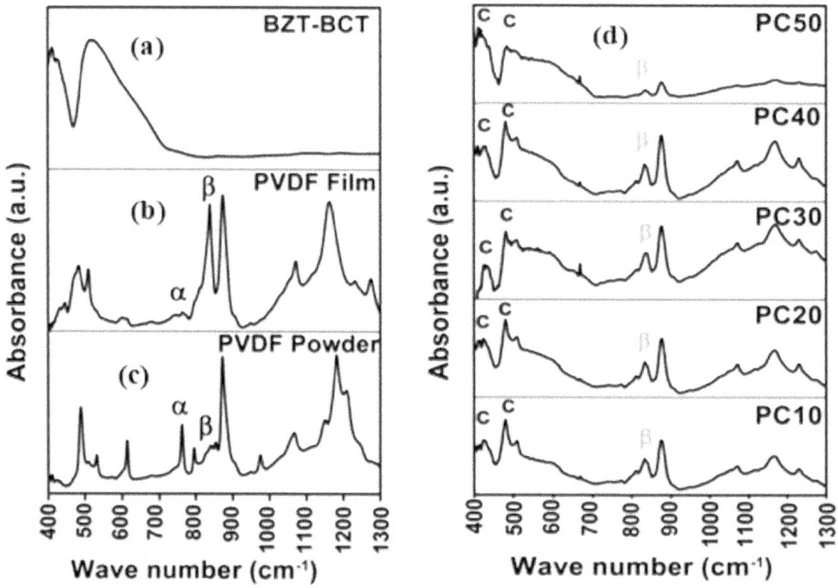

FIGURE 6.3 FTIR spectra of (a) BZT-BCT; (b) PVDF film; (c) PVDF powder; (d) composites.

6.3.3 MORPHOLOGY

SEM images of the composite samples are shown in Figure 6.4. Figure 6.4(a) represents the HR SEM image of BZT-BCT ceramic power and it can be clearly observed that particles are of almost spherical shape and size ≤ 1 μm. Figure 6.4(b) shows the SEM image of pure PVDF film, which is very smooth. SEM images of the surface of PVDF-(BZT-BCT) composite are shown in Figure 6.4(b)–(g) in increasing order of ceramic content. It can be clearly observed that the compactness of ceramic particles increases with the increment of ceramic content in the PVDF matrix. It also indicates that the ceramic particles are well dispersed in the polymer matrix. Optical images of the films are represented in the inset of corresponding SEM images.

6.3.4 DIELECTRIC PROPERTIES

Frequency-dependent dielectric constants and tanδ are given in Figures 6.5(a) and (b). The nature of the permittivity curves, obtained for all the

FIGURE 6.4 SEM images of ceramic particle and composite films.

compositions, is similar to that of pure PVDF film. All the dielectric curves show a high dielectric constant at a lower frequency and rapid decrement up to 1 kHz, whereas there is a slight decrement in the range of 10^5 to 10^6 Hz. Dielectric constant in different frequency ranges depends upon different polarizations such as ionic, dipolar, electronic, and interfacial polarization [16]. With the increase of frequency, the contributions of both dipolar polarization and interfacial polarization decrease; that is why the dielectric constant for all the samples shows a rapid reduction up to 1 kHz [16, 20]. Interfacial polarization increases with the addition of ceramic particles into the PVDF matrix; which causes overall higher permittivity [22]. Dielectric constant of PC50 is about 19, which is almost two times that of PVDF(10.2) at 1 kHz. This may be due to the strong interfacial polarization at the interface of PVDF and BZT-BCT [22]. Figure 6.5(c) represents frequency-dependent dielectric constant and tanδ of BZT-BCT, which is having ε' as 2,508 at 1 kHz frequency and well matches with the literature [14]. Loss tangent of PVDF and BZT-BCT are 0.04 and 0.02 [14] at 1 kHz frequency, respectively, but that of composites show almost equal tanδ nearly about 0.04.

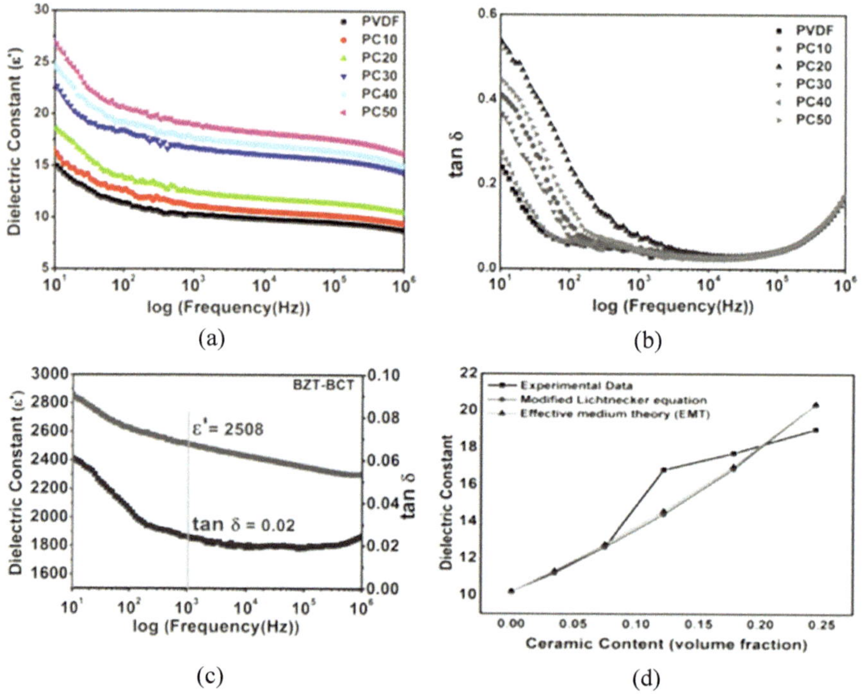

FIGURE 6.5 Dependence of (a) dielectric constant of composites; (b) tanδ of composites; (c) dielectric constant and tanδ of BZT-BCT on frequency at room temperature; (d) theoretical variation of dielectric constant with ceramic content.

The fast decreasing nature of tanδ curves of composites at lower frequency region may be attributed to Maxwell-Wagner relaxation and increasing after frequency >10^5 corresponds to α relaxation of PVDF (glass transition relaxation) [17, 24]. All the tanδ curves show common minimum in the frequency range (10^4 Hz) < f < (10^5 Hz). In this frequency range, maximum numbers of dipoles oscillate themselves according to applied field frequency, which causes minimum dielectric loss [7]. Figure 6.5(d) shows the experimental variation trend of dielectric constant (at 1 kHz) with an increase in ceramic content, and the same is fitted with theoretical variation trends given by the modified Lichtnecker equation (Eqn. (3)) and effective medium theory (EMT) (Eqn. (4)) [10].

$$\log \varepsilon = \log \varepsilon_m + v_f(1-n)\log\left(\frac{\varepsilon_i}{\varepsilon_m}\right) \tag{3}$$

$$\varepsilon = \varepsilon_m\left[1+\frac{v_f(\varepsilon_i-\varepsilon_m)}{\varepsilon_m+n(1-v_f)(\varepsilon_i-\varepsilon_m)}\right] \tag{4}$$

where; ε_i and ε_m are the dielectric constants of filler and matrix, respectively. v_f is the filler volume fraction, and n is the ceramic morphology fitting factor. After fitting, the values of *n* were found to be 0.32 from the EMT model and 0.48 from the Lichtnecker equation. Therefore, the close agreement between the experimental and theoretical values can be attributed to the morphology of the dispersed particles (almost spherical particles) [10].

6.3.5 FERROELECTRIC PROPERTIES

The ferroelectric hysteresis (PE) loop for all the composite samples is shown in Figure 6.6(a). Unlike BZT-BCT [14], PE loops obtained for all the composites along with pure PVDF are saturation less [25]. The sudden increment of remnant polarization (P_r) of PVDF to that of PC10 and further increase with the increase of ceramic content can be noticed in Figure 6.6(b). This indicates the contribution of ceramic towards the effective P_r of the composite due to increment in dielectric constant [7]. The coercive field (E_c) of the composites is almost constant and slightly higher than that of pure PVDF. In Figure 6.6(d), the area within the PE loop (W_2) represents the non-recoverable energy density or ferroelectric loss energy density, and this depends on both P_r and E_c [7]. The area W_1 represents recoverable energy density, and the sum ($W_1 + W_2$) represents the total storage energy density for the material. So, the efficiency of storage energy density is given by $W_1/(W_1 + W_2)$. W_1 can be calculated using the following equation [20]:

$$W_1 = \int_{P_r}^{P_{max}} E\,dP \tag{5}$$

where; E, P, and P_{max} represent the applied electric field, polarization, and maximum polarization, respectively. It can be clearly observed that for the constant maximum field (70 kV/cm) and frequency (10 Hz), the loop area increases with the increment of ceramic content, so the efficiency of storage energy changes from 90.12% to 64.38% for a ceramic fraction of x = 0 to 0.5 (see Figure 6.6(c)). The addition of particles with a high dielectric constant distorts the electric field distribution, which causes ferroelectric loss resulting in lesser storage efficiency [17, 23]. But at the same time, recoverable storage energy density increases from 2.1 mJ/cm3 to 5.51 mJ/cm3 for x = 0 to 0.5. This may be attributed to the trapped dipoles that are not altering their polarity even after the alteration of the applied electric field [25]. So, the composite with higher ceramic content can be considered for energy harvesting applications.

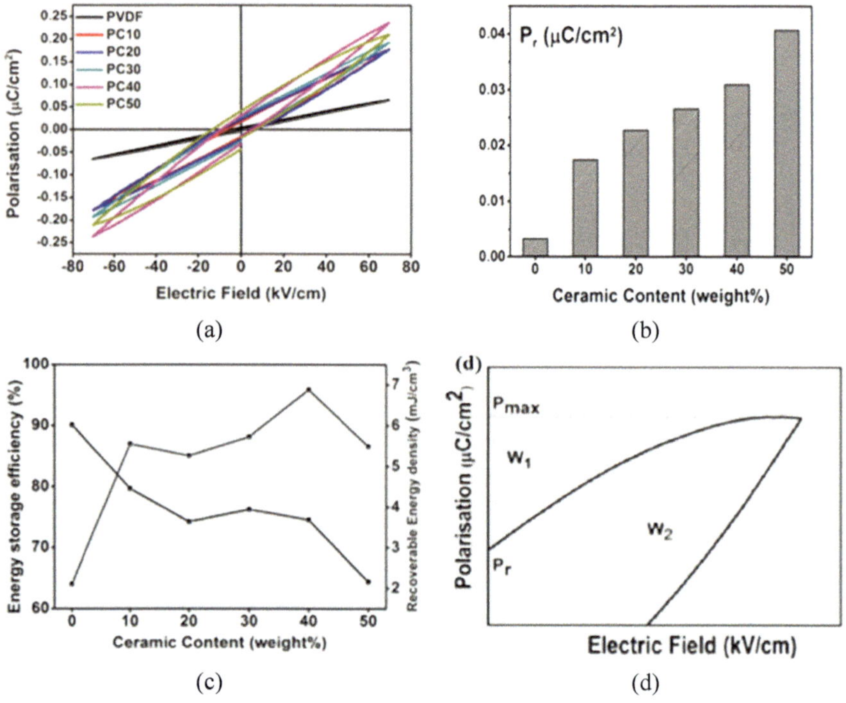

FIGURE 6.6 (a) Ferroelectric hysteresis loop of composites; (b) variation of P_r with ceramic content; (c) variation of energy storage efficiency and recoverable energy density with ceramic content; (d) area representing different energy densities.

6.4 CONCLUSION

Polymer ceramic composites were prepared by adding BZT-BCT ceramic into PVDF matrix. Films were prepared by solvent casting method and then hot pressed. Formation of β-phase of PVDF in the composite films was verified by XRD and Fraction of β-phase was estimated from FTIR spectroscopy. Fraction of β-phase is somewhat reduced in composites than that of pure PVDF film. The morphology of the films was studied by SEM. The dielectric constant of the composite films increases with BZT-BCT content. PC50 has the highest dielectric constant, i.e., at 1 kHz, ε' of PC50 is 19, which is almost double of ε' of PVDF (10.2). For all the samples, the dissipation factor was found to be < 0.1. To study the energy storage property, the ferroelectric PE loop was traced. Remnant polarization increases with an increase in ceramic content. The PE loop of the composites has more loop area, indicating lossy nature. Finally, it can be concluded that the addition of BZT-BCT to the PVDF matrix enhances dielectric and ferroelectric properties, and PVDF-(BZT-BCT) composite can be preferred over pure PVDF film for energy harvesting applications.

ACKNOWLEDGMENTS

The corresponding author is pleased to acknowledge the Department of Science and Technology, Government of India, New Delhi, for providing financial support through SERB research grant No. ECR/2017/000281. Authors also acknowledge DST-SAIF, Kochi, for XRD characterization, and SAIF, IIT Madras for SEM.

KEYWORDS

- **dielectric properties**
- **Fourier transform infrared**
- **glass transition relaxation**
- **poly(vinylidene fluoride)**
- **polymer ceramic composite**
- **scanning electron microscope**
- **storage energy density**

REFERENCES

1. Niu, Y., Yu, K., Bai, Y., & Wang, H., (2015). Enhanced dielectric performance of $BaTiO_3$/PVDF composites prepared by modified process for energy storage applications. *IEEE Transactions on Ultrasonics, Ferroelectrics, and Frequency Control, 62*, 108. https://doi.org/10.1109/TUFFC.2014.006666.
2. Kalani, S., Kohandani, R., & Bagherzadeh, R., (2020). Flexible electrospun PVDF–$BaTiO_3$ hybrid structure pressure sensor with enhanced efficiency. *RSC Advances, 10*(58), 35090–35098. https://doi.org/10.1039/D0RA05675H.
3. Guo, M., Jiang, J., Shen, Z., Lin, Y., Nan, C. W., & Shen, Y., (2019). High-energy-Density ferroelectric polymer nanocomposites for capacitive energy storage: Enhanced breakdown strength and improved discharge efficiency. *Materials Today, 29*, 49–67. https://doi.org/10.1016/j.mattod.2019.04.015.
4. Wankhade, S. H., Tiwari, S., Gaur, A., & Maiti, P., (2020). PVDF–PZT nanohybrid based nanogenerator for energy harvesting applications. *Energy Reports, 6*, 358–364. https://doi.org/10.1016/j.egyr.2020.02.003.
5. Alizadeh-Osgouei, M., Li, Y., & Wen, C., (2019). A comprehensive review of biodegradable synthetic polymer-ceramic composites and their manufacture for biomedical applications. *Bioactive Materials, 4*, 22–36. https://doi.org/10.1016/j.bioactmat.2018.11.003.
6. Ruan, L., Yao, X., Chang, Y., Zhou, L., Qin, G., & Zhang, X., (2018). Properties and applications of the β phase poly(vinylidene fluoride). *Polymers, 10*(3), 228. https://doi.org/10.3390/polym10030228.
7. Muduli, S. P., Parida, S., Rout, S. K., Rajput, S., & Kar, M., (2019). Effect of hot press temperature on β-phase, dielectric and ferroelectric properties of solvent casted poly(vinylidene fluoride) films. *Materials Research Express, 6*, 095306. https://doi.org/10.1088/2053-1591/ab2d85.
8. Nurbaya, Z., Wahid, M. H., Rozana, M. D., Gan, W. C., Majid, W. H., Alrokayan, S. A., Khan, H. A., & Rusop, M., (2016). Preparation of PVDF-TrFE layer-based bilayer composite $PbTiO_3$/PVDF-TrFE films for MIM capacitor. *Transactions of the IMF, 94*, 187. http: //dx.doi.org/10.1080/00202967.2016.1180809.
9. Seema, A., Dayas, K. R., & Varghese, J. M., (2007). PVDF-PZT-5H composites prepared by hot press and tape casting techniques. *Journal of Applied Polymer Science, 106*, 146. https://doi.org/10.1002/app.26673.
10. Jain, A., Prashant, K. J., Sharma, A. K., Jain, A., & Rashmi, P. N., (2015). Dielectric and piezoelectric properties of PVDF/PZT composites: A review. *Polymer Engineering & Science, 55*, 1589. https://doi.org/10.1002/pen.24088.
11. Singh, P., Borkar, H., Singh, B. P., Singh, V. N., & Kumar, A., (2014). Ferroelectric polymer-ceramic composite thick films for energy storage applications. *AIP Advances, 4*(8), 087117. http: //dx.doi.org/10.1063/1.4892961.
12. Panda, P. K., & Sahoo, B., (2015). PZT to lead free piezo ceramics: A review. *Ferroelectrics, 474*, 128. https://doi.org/10.1080/00150193.2015.997146.
13. Mishra, P., & Kumar, P., (2012). Effect of sintering temperature on dielectric, piezoelectric and ferroelectric properties of BZT–BCT 50/50 ceramics. *Journal of Alloys and Compounds, 545*, 210. https://doi.org/10.1016/j.jallcom.2012.08.017.

14. Coondoo, I., Panwar, N., Amorín, H., Alguero, M., & Kholkin, A. L., (2013). Synthesis and characterization of lead-free 0.5 Ba $(Zr_{0.2}Ti_{0.8})$ O_3-0.5 $(Ba_{0.7}Ca_{0.3})$ TiO_3 ceramic. *Journal of Applied Physics, 113*, 214107. https://doi.org/10.1063/1.4808338.
15. Yan, X., Zheng, M., Gao, X., Zhu, M., & Hou, Y., (2020). High-performance lead-free ferroelectric BZT–BCT and its application in energy fields. *Journal of Materials Chemistry, C.* https://doi.org/10.1039/D0TC03461D.
16. Pandey, B. K., Kumar, A., Chandra, K. P., Kulkarni, A. R., Jayaswal, S. K., & Prasad, K., (2018). Electrical properties of 0–3 0.5 $(Ba_{0.7}$ $Ca_{0.3})$ TiO_3–0.5 Ba $(Zr_{0.2}Ti_{0.8})O_3$/PVDF nanocomposites. *Journal of Advanced Dielectrics, 8*(04), 1850027. https://doi.org/10.1142/S2010135X18500273.
17. Chi, Q., Liu, G., Zhang, C., Cui, Y., Wang, X., & Lei, Q., (2018). Microstructure and dielectric properties of BZT-BCT/PVDF nanocomposites. *Results in Physics, 8*, 391–396. https://doi.org/10.1016/j.rinp.2017.12.052.
18. Li, Y. C., Li, R. K., & Tjong, S. C., (2010). Frequency and temperature dependences of dielectric dispersion and electrical properties of polyvinylidene fluoride/expanded graphite composites. *Journal of Nanomaterials*. https://doi.org/10.1155/2010/261748.
19. Mishra, P., & Kumar, P., (2015). Structural, dielectric, and optical properties of [(BZT–BCT)-(epoxy-CCTO)] composites. *Ceramics International, 41*, 2727. https://doi.org/10.1016/j.ceramint.2014.10.087.
20. Muduli, S. P., Parida, S., Nayak, S., & Rout, S. K., (2020). Effect of graphene oxide loading on ferroelectric and dielectric properties of hot-pressed poly(vinylidene fluoride) matrix composite film. *Polymer Composites, 41*, 2855–2865. https://doi.org/10.1002/pc.25581.
21. Abdelhamid, E. H., Jayakumar, O. D., Kotari, V., Mandal, B. P., Rao, R., Naik, V. M., Naik, R., & Tyagi, A. K., (2016). Multiferroic PVDF–Fe_3O_4 hybrid films with reduced graphene oxide and ZnO nanofillers. *RSC Advances, 6*, 20089. https://doi.org/10.1039/C5RA26983K.
22. Martins, P., Lopes, A. C., & Lanceros-Mendez, S., (2014). Electroactive phases of poly(vinylidene fluoride): Determination, processing and applications. *Progress in Polymer Science*, *39*, 683. https://doi.org/10.1016/j.progpolymsci.2013.07.006.
23. Liu, S., Xue, S., Xiu, S., Shen, B., & Zhai, J., (2016). Surface-modified Ba $(Zr_{0.3}$ $Ti_{0.7})$ O_3 nanofibers by polyvinylpyrrolidone filler for poly(vinylidene fluoride) composites with enhanced dielectric constant and energy storage density. *Scientific Reports, 6*, 26198. https://doi.org/10.1038/srep26198.
24. Fu, J., Hou, Y., Zheng, M., Wei, Q., Zhu, M., & Yan, H., (2015). Improving dielectric properties of PVDF composites by employing surface-modified strong polarized $BaTiO_3$ particles derived by molten salt method. *ACS Applied Materials & Interfaces, 7*, 24480. https://doi.org/10.1021/acsami.5b05344.
25. Riquelme, S. A., & Ramam, K., (2019). Dielectric and piezoelectric properties of lead free BZT-BCT/PVDF flexible composites for electronic applications. *Materials Research Express, 6*(11), 116331. https://doi.org/10.1088/2053-1591/ab522c.

CHAPTER 7

Structural and Frequency-Dependent Electrical Properties of Lead-Free Na_2 Ba_2 La_2 W_2 Ti_4 Nb_4 O_{30} Ceramics

S. DEVI,[1] S. BEHERA,[1] and T. SAHU[2]

[1]*Department of Physics, Centurion University of Technology and Management, Odisha, India,*
E-mail: saubhagyalaxmi.behera@cutm.ac.in (S. Behera)

[2]*School of Physics, Sambalpur University, JyotiViahr, Burla–768019, Odisha, India*

ABSTRACT

The eco-friendly tungsten bronze (TB)-type ceramic oxide, Na_2 Ba_2 La_2 W_2 Ti_4 Nb_4 O_{30} was synthesized by a solid solution technique. The phase and crystal structure of the compound is studied from XRD analysis. The microstructure of the material is investigated through scanning electron microscope (SEM) techniques. The ferroelectric nature of the compound is studied from the temperature variation of dielectric parameters with the ferroelectric transition at 310C, which may be later confirmed by a polarization study. In the impedance spectroscopy analysis, the existence of single semicircles in the Nyquist plot manifests the presence of grain conduction within the reported temperature range. The type of charge carriers in the ceramic is studied from the ac, and dc conductivity is plotted against temperature. The variation of ac conductivity with frequency at different temperatures supports the hopping of charge carriers in the material.

Advances in Diverse Applications of Polymer Composites: Synthesis, Application, and Characterization.
Suji Mary Zachariah, Yang Weimin, Maciej Jaroszewski, & Sabu Thomas (Eds.)

7.1 INTRODUCTION

Ferroelectric ceramics with tungsten bronze (TB) structure is the popularly studied group of dielectrics, just following the perovskites. Commonly, most of the ferroelectric oxides possess BO_6 oxygen octahedral structure irrespective of their crystal structure, transition temperature, remnant polarization, electrical, and mechanical properties. The general formula $(A_1)_2(A_2)_4(C)_4(B_1)_2(B_2)_8O_{30}$ for TB structure contains 10 distorted octahedral with three different corners (2tetragonal*A*1, 4pentagonal*A*2, and 4triangle*C*) accommodates various A, B, and C cations in their sites [1, 2]. The beauty of this family of oxides is that useful properties can be modified by a variety of cation substitution in different sites [3]. From various cations of the above formula, A_1 and A_2 sites are accommodated by metallic cations like K+, Na+, Sr^{2+}, Ba^{2+}, Ca^{2+}, Pb^{2+} and the rare earths.; the B cations are occupied by heavy metals like W^{6+}, Nb^{5+}, Ta^{5+}, Zr^{4+}, Ti^{4+}, Sb^{5+}; and the C sites by Li^{+}, Be^{2+}, Mg^{2}. The ferroelectricity in tungsten bronze structure are due to interaction between coordination status of the octahedral B-site cations [4, 5]. Usually, the smallest C cations are absent in the chemical formula. When six cations are placed in the A site of the formula $(A_1)_4(A_2)_2(B)_{10}O_{30}$, the structure is named as 'filled' TB structure; unless it is referred as 'unfilled' TB structure [6]. In view of the above, a good no of filled tungsten bronze ceramic with lead at A site are investigated earlier due to their distinctive structure, and various interesting properties [7–10]. At the same time, PbO being a raw materials for the above lead-based tungsten bronze oxides, is highly toxic and its toxicity further increases due to its evaporation at high-temperature calcination and sintering, creating environmental pollution. Hence replacement of lead from dielectrics without much more affecting its properties is a challenge for the material scientist [11]. In view of the above various rare-earth ions substituted with lead-free tungsten bronze niobates have been reported [12, 13]. From the literature survey, it is confirmed that the substitution of barium by lead in the A site of rare-earth-based niobates is not explored; hence we systematically synthesized and characterized $Na_2Ba_2La_2W_2Ti_4Nb_4O_{30}$ and presented its various modified properties.

7.2 EXPERIMENTAL

The polycrystalline ceramic $Na_2Ba_2La_2W_2Ti_4Nb_4O_{30}$ (abbreviated as NBLN) was fabricated by a cost-effective solid-state technique followed by high calcination and sintering temperature. The high purity stoichiometric precursors: Na_2CO_3 (99%, M/ss.d.Finechem.Ltd.), $BaCO_3$ (99.9%, E.MerckLtd., India), TiO_2, WO_3, Ta_2O_5 (99%, M/sLobaChemie Pvt. Ltd., India), and La_2O_3 (99.9%, Indian RareEarth Ltd., India) are weighed and converted into a homogeneous mixture by grinding it manually for three hours in dry and wet (methanol) medium. The mixture is calcined at 1,150°C (on the basis of the calcination temperature of the compound of a similar niobate family) for 4 hrs in an air medium using a high purity alumina crucible. The compound formation is checked from XRD; otherwise, the above process is repeated. The powder diffractometer (Rigakuminiflex) with Cukα radiation ($\lambda = 1.5405$ Å) is used to check the quality and phase of the material at room temperature taken over a range of Bragg's angle (2θ) ($200 \leq 2\theta \leq 800$) at a scanning rate of 3 deg/min. The calcined powders were grounded and mixed with organic binder PVA(Polyvinyl alcohol) to make cylindrical pellets, of 10 mm diameter and 1–2 mm thickness at the pressure of 4.5×106 Nm^{-2}. The process of sintering at an optimized temperature of 1,200°C for 4 hrs yields dense ceramics evaporating the binder at elevated temperature. Both the plane surfaces of a sintered pellet were polished and painted with gold by sputtering technique for better resolution. The microstructure was studied by scanning electron microscope (SEM; JEOLJSM-5800). The sintered pellets were polished and painted with conductive silver and again heated at 160°C for 1 h to make it free from moisture (if any) for electrical measurement. The dielectric and electrical properties of the compound is measured with a computer-controlled Hioki 3532 LCR Hitester in the frequency range of 102–106 Hz at various temperatures (29–500°C).

7.3 RESULT AND DISCUSSION

7.3.1 STRUCTURE/MICROSTRUCTURE

The fine and well-defined peaks observed from the room temperature XRD pattern of the powdered sample, which don't overlap with the peaks of the starting materials (Figure 7.1) confirm the formation of the compound in a single phase. A computer program package, "POWDMULT" is used to index all the peaks in various crystal systems and cell configurations and based on the least-squares refinement on the minimum value between the observed and calculated interplanar spacing, orthorhombic unit cell was chosen [14]. The lattice parameters of the compound after refinement are, a = 21.4310 Å, b = 17.766 Å and c = 3.7830 Å which has well agreement with reported compound [13]. Scherrer's equation, P = (0.89λ/β1/2Cosθ), where λ = 1.5405 Å and β1/2 = full width at half maximum (in radians) gives the crystallite size of the powder sample 16 nm using the spreading of reflection peaks.

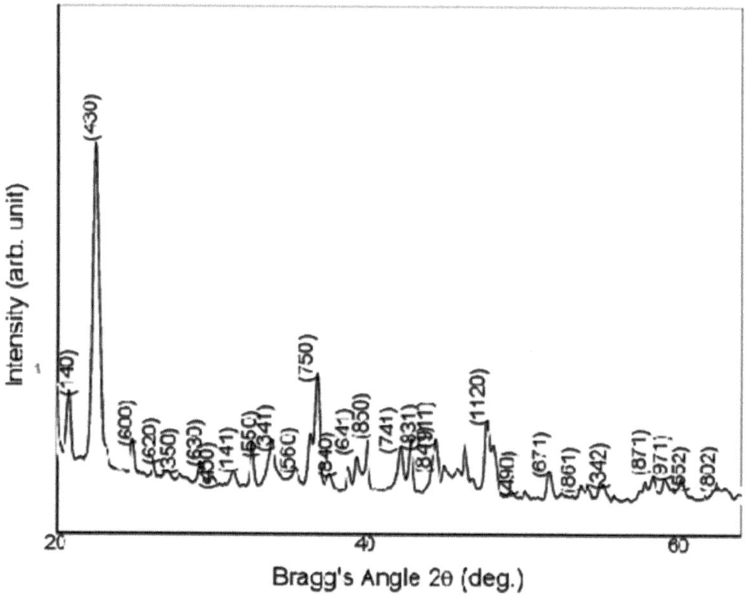

FIGURE 7.1 X-ray diffraction pattern of NBLN at room temperature.

The room temperature SEM micrograph (Figure 7.2) of the compound at various magnifications shows polycrystalline nature of the material. The rod-shaped grains are non-homogeneously distributed in the sample surface with average grain lying between 3 μm and 12 μm.

7.3.2 DIELECTRIC STUDY

The temperature dependence of the relative dielectric constant and dielectric loss at different frequencies are shown in Figures 7.3(a) and (b). Initially, the dielectric constant rises with increasing temperature due to the rapid motion of the cations, attains peak (maximum) value at 583 K, and then decreases up to 608 K, then after increases above 608 K. The dipolar

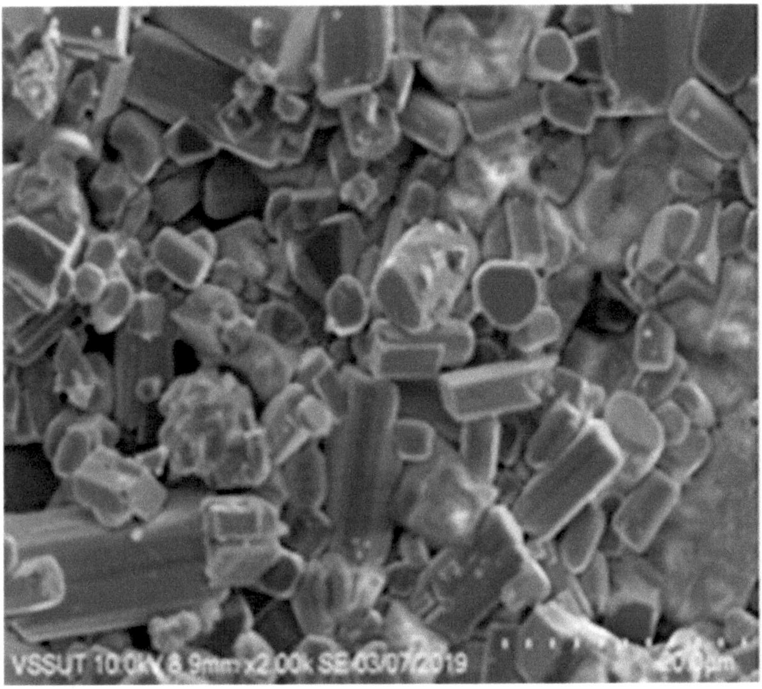

FIGURE 7.2 SEM micrograph of NBLN.

ordering, electron-phonon interaction enhances the dielectric constant up to transition temperature, and beyond Tc, εr decreases due to paraelectric behavior. Again, at much higher temperature above Tc, thermally activated mobile charge carriers and space charge polarization further increases the dielectric constant. High-temperature sintering creates single or double ion oxygen vacancy, which are known as thermally activated space charge and produces valency fluctuation at B sites between Nb^{5+}–Nb^{4+} and Ti^{4+}–Ti^{3+} ions [16]. The applied electric field also accelerates the electron along its direction between Nb^{5+}–Nb^{4+} and Ti^{4+}–Ti^{3+} at octahedral sites again contributes to the dielectric constant in addition to oxygen vacancy [17]. At lower frequency (10 kHz), the accumulation of space charge at grain boundaries makes the dielectric constant high [18]. The removal of space charge polarization with increasing frequency reduces the dielectric constant. The non-Relaxor behavior of the material is explained from the frequency independence of transition temperature. The dielectric constant (ε_{max}) of the compound at Tc for frequencies at 10,100 kHz and 1 MHz are found to be 638,316 and 206, respectively, which are greater than that lead-based compound reported by our group [10]. Substitution of barium in place of lead shifts the phase transition temperature from 536 K to 586 K. The dielectric loss (tanδ) varies with temperature in a similar way as dielectric constant with peak closer to Tc (573 K), which is a normal behavior of tungsten bronze type compounds [19]. The decrease in the ferroelectric domain wall contribution and accumulation of free charge carriers near the boundary within the potential barrier at high temperatures rises the magnitude of tanδ. The observed peak value of dielectric loss at Tc for frequencies at 10,100 kHz and 1 MHz, are 2.18, 0.72, and 0.39, respectively, which are similar to those observed [10].

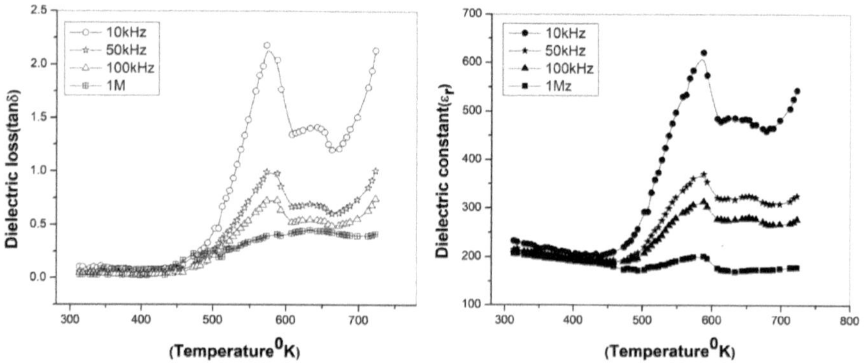

FIGURE 7.3 Variation of ε_r and tanδ of NBLN as a function of temperature at 10 kHz, 100 kHz, and 1 MHz, respectively.

The slight decrease of dielectric constants above the peaks gives the phenomenon of diffuse phase transition in the compound. The degree of diffuseness at the peak is calculated using a general expression:

$$\text{In}\ (1/\varepsilon_r - 1/\varepsilon_{\max}) = \text{In}\ (T - T_c) + constant \quad (1)$$

where ε_r is the dielectric constant, $\varepsilon_{\max}$ is the maximum dielectric constant, T_c is the transition temperature.

The value of diffusivity (γ), calculated from the slope of $\ln(1/\varepsilon_r - 1/\varepsilon_{\max})$ vs. $\ln(T - T_c)$ plots (Figure 7.4), at 10 kHz was found to be 1.22. The diffuseness parameter γ with intermediate value between 1 and 2 (1 for ideal ferroelectric and 2 for relax or ferroelectric) confirms the phase transition of diffused type in the material [20]. Most of the niobate-based

TB compounds exhibit diffuse phase transition as a result of the disorder arrangement of cations.

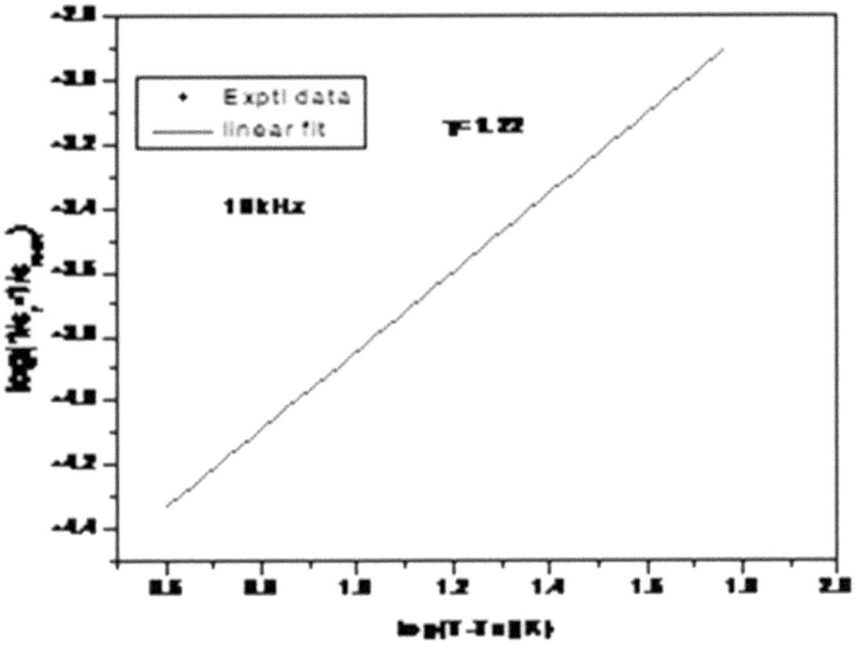

FIGURE 7.4 Variation of $\ln[1/\varepsilon_r - 1/\varepsilon_{max}]$ with $\ln(T - T_c)$ at 10 kHz of NBLN.

7.3.3 IMPEDANCE STUDY: ELECTRICAL PROPERTIES

For better clarification about the electrical behavior of ionic and polycrystalline crystals impedance spectroscopy is mostly used. The inter-grain, intra-grain, and other electrode effects are usually responsible for the electrical properties of ferroelectric materials. The contributions due to grains, grain boundaries and interface can be easily distinguished by this technique. Sometimes it is hard to interpret the response due to the long-range motion of the charges or dipole relaxation in the material. Some important data having both real (resistive) and imaginary (reactive) components are provided by the measurements of impedance and related parameters of the materials. Using the below-given formulas, we can calculate these components:

$$\text{Complex impedance } Z(\omega) = Z' - jZ'' = R_s - \frac{j}{\omega C_s} \quad (2)$$

$$\text{Complex electrical modulus } M(\omega) = \frac{1}{\varepsilon(\omega)} = M' + j\,M'' = j\,\omega CoZ \quad (3)$$

$$\text{Complex admittance } Y^* = Y' + jY'' = j\omega C_o\varepsilon^* = (R_P)^{-1} + j\omega C_p \quad (4)$$

$$\text{Complex permittivity} \varepsilon^* = \varepsilon' - j\varepsilon'' \quad (5)$$

where; $\omega = 2\,\pi f$ is the angular frequency; C_0 is the geometrical capacitance; $j = \sqrt{-1}$ and subscripts 'p' and 's' are parallel and series circuit components, respectively.

The graph between Z′ and frequency at selected temperatures is given in Figure 7.5(a). The lowering of Z′ value at high frequency and temperature range suggests the increase of ac conductivity in the material. At low frequencies oxygen vacancies have a large contribution to the conductivity, however, when the frequency becomes high the Z′ values overlap for all the temperature indicating the inability of space charges(oxygen vacancies) to respond to the high-frequency fields. The signature of NTCR behavior like semiconductor is suggested from the lowering of Z′ with temperature.

The loss spectrum Z″ verses frequency is shown in Figure 7.5(b). Appearance of peaks at a particular frequency (called relaxation frequency) for all the given temperature is noticed from the graph. The peak values decrease with increasing measuring temperature. The shifting of the peak frequencies towards high with rise of temperature indicates increase in relaxation and decrease of barrier properties in the material. The width of the peaks increases with increasing temperature denoting diffuse phase transition and thermally activated electrical relaxation phenomenon in the material [21]. The defects or vacancies created at higher temperature and the fixed charges or electrons at low temperatures contributes towards relaxation process [22]. At higher temperature, the slopes of the graph remain the same irrespective of temperature while the low-frequency slopes vary significantly with temperature indicating two distinct dispersion mechanisms in the compound.

Figure 7.6 displays the Cole-Cole plot of the material at a selected temperature, which presents large arcs at intermediate frequencies. The curve fitting was done by ZSIMP-WIN (version 2) software. The curve was best fitted with an equivalent circuit containing a parallel combination of (CQR) at low temperature. The formula for the above plot is written by:

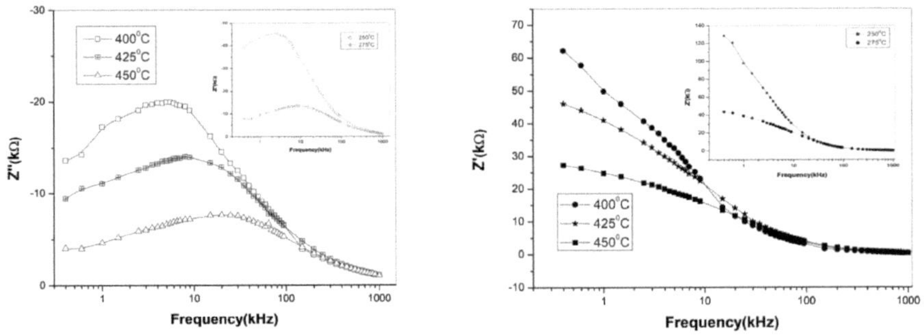

FIGURE 7.5 Variation of Z′ and Z″ of NBLN with frequency at selected temperatures.

$$Z^*(\omega) = R/\{1 + (j\,\omega/\omega_0)^{1-n}\} \tag{6}$$

where; exponent $n \rightarrow 0$ (i.e., $1 - n = 1$).

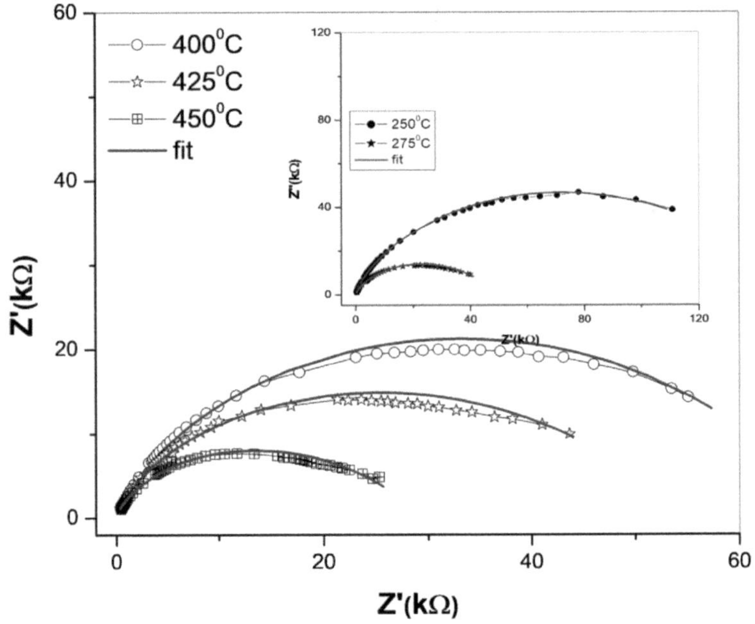

FIGURE 7.6 Variation of Z″ with Z′ of NBLN at different temperatures.

And this equation contributes to the classical Debye's relation [23]. The impedance plots show single semicircular arches with center below

Z' axis for all the selected temperatures, which means that the system has an access to the single conduction mechanism with non-Debye type. This is because of the grain size of the material or mobility of polarons induced by the increase of temperature. The intercept of arcs on the Z' axis determines the value of grain resistance (R_g), which reduces with enhancing the temperature indicating the presence of thermal conduction mechanism as well as NTCR behavior of the material [23, 24].

7.3.4 DC CONDUCTIVITY

The electrical conductivity(dc) of the compound is independent of frequency and the free charge carriers are responsible for this. The formula for this conductivity is given by relation:

$$\sigma_{dc} = t/R_bA \tag{7}$$

where; R_b is the bulk resistance; 't' the thickness; and 'A' is the surface area of the sample, respectively. The temperature variation of dc conductivity is shown in Figure 7.7. The dc conductivity increases with rising temperature and follows the Arrhenius relation:

$$\sigma_{dc} = \sigma_o e^{-\frac{E_a}{K_B T}} \tag{8}$$

The slope of the above graph determines the activation energy and was found to be 0.25 eV.

7.3.5 AC CONDUCTIVITY

The plot of σ_{ac} verses frequency at selected temperature is reflected in Figure 7.8. Considering that the dielectric loss in the above temperature range contributes to conductivity, it is given by the formula, $\sigma(\omega) = \omega\varepsilon 0\varepsilon''$ where, σ is the ac conductivity. The response of ac conductivity towards frequency is divided into two parts for both the low-frequency and high-frequency regions for all temperatures. In the lower frequency region, σ_{ac} does not change with frequency and in higher frequency, it rises with frequency. So, σ_{dc} is calculated by extrapolating the frequency-independent part. The ac conductivity obtained at higher frequencies arises by hopping of charge carriers from long-range to short-range ionic motion [25, 26]. The dispersion decreases with increasing temperature, which confirms the absence of electrode polarization in the compound. At low frequencies, the small density of oxygen vacancies reduces the drift velocity of the charge carriers; hence, the conductivity is less. All the conductivity curves overlap with each other at high frequencies causing temperature and frequency independent dispersion in the compound. The ac conductivity obeys Jonscher's power law:

$$\sigma_{ac} = \sigma_{dc} + A\omega^n \tag{9}$$

where; σ_{dc} is DC conductivity, the frequency exponent term 'n' varies between en 1 and 2; and 'A' gives the polarizability strength. The experimental data are fitted with the above equation for two temperatures 275°C and 450°C, respectively (Figure 7.9). The estimated value of *n* is 0.715 and 0.66 for the sample at temperatures 275C and 450°C ($0 < n < 1$), respectively.

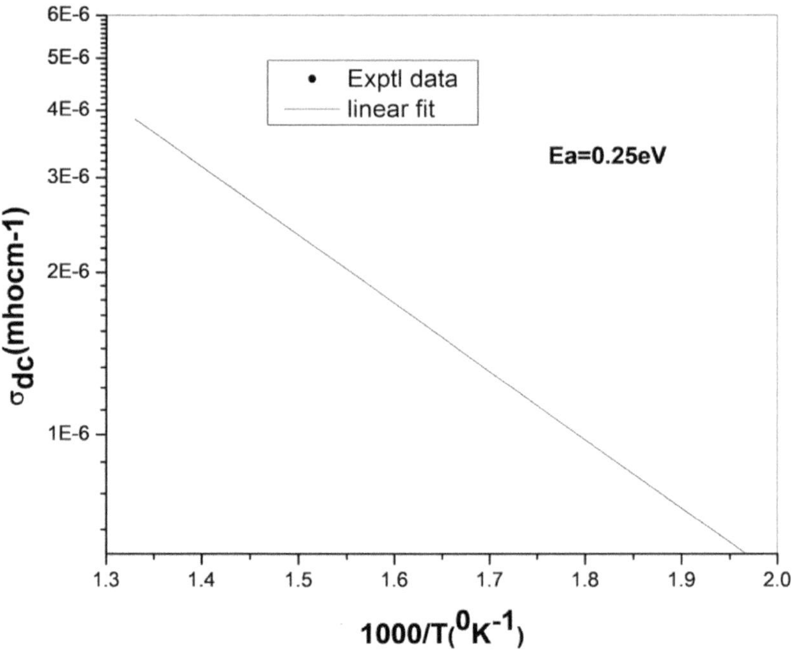

FIGURE 7.7 Variation of DC conductivity with inverse of absolute temperature of NBLN.

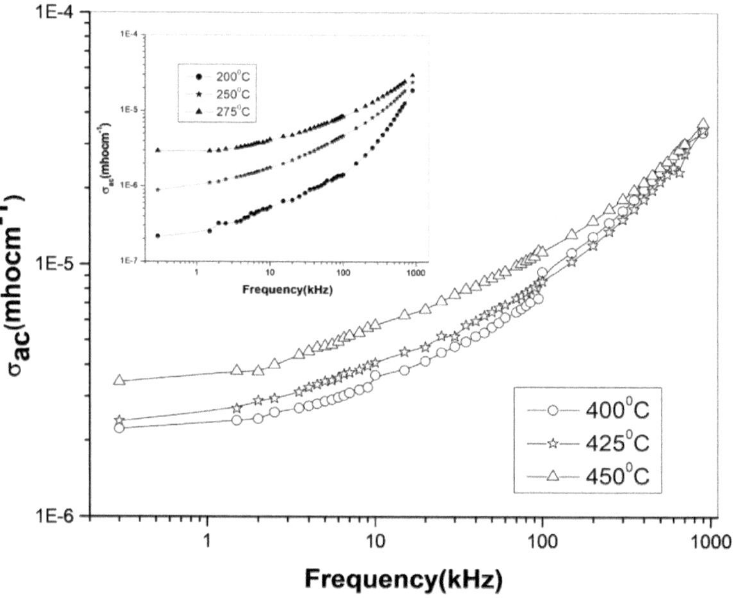

FIGURE 7.8 Variation of AC conductivity with frequency of NBLN at different temperatures.

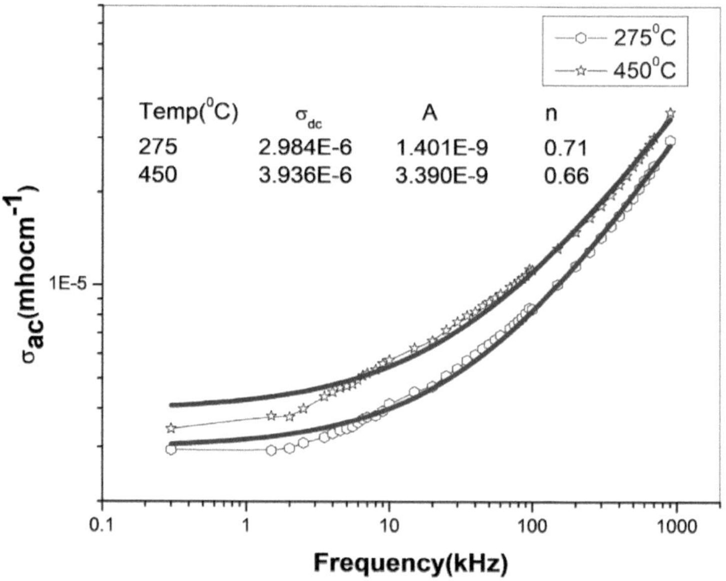

FIGURE 7.9 AC fitting curve for NBLN at different temperatures.

7.4 CONCLUSION

The tungsten bronze structured ceramic $Na_2Ba_2La_2W_2Ti_4Nb_4O_{30}$ was synthesized by mixed oxide technique at a calcination temperature of 1,150°C. The single-phase material with an orthorhombic crystal structure is confirmed from XRD. The diffused ferroelectric phase transition with a high Tc is studied from temperature-dependent dielectric parameters. The spectroscopic analysis technique gives information regarding conduction, relaxation process, and semiconducting behavior. The single semicircular arcs in the Nyquist plot revealed the contribution of grain effects on the electrical properties. The translational motion of mobile charge carriers is studied from the ac conductivity spectrum.

KEYWORDS

- **AC conductivity**
- **ceramics**
- **impedance spectroscopy**
- **niobates**
- **polycrystalline**
- **scanning electron microscope**
- **solid-state reaction**

REFERENCES

1. Xu, Y., (1999). *Ferroelectric Materials and their Applications* (p. 247). Elsevier, North-Holland, Amsterdam.
2. Jamieson, P. B., Abrahams, S. C., & Bernstein, J. L., (1969). *J. Chem. Phys., 48*, 5048–5057.
3. Fang, L., Meng, S. S., & Hu, C. Z., (2007). *J. Alloys Compd., 429*, 280–284.
4. Xie, R. J., & Akimune, Y., (2002). *J. Mater. Chem., 12*, 3156–3161.
5. Wei, L. L., Yang, Z. P., Chang, Y. F., & Gu, R., (2008). *J. Am. Ceram. Soc., 91*, 1077–1082.
6. Lin, K., Rong, Y. C., Wu, H., Huang, Q. Z., You, L., Ren, Y., Fan, L. L., et al., (2014). *Inorg Chem., 53*, 9174–9180.
7. Das, P. R., Parida, B. N., Padhee, R., & Choudhary, R. N. P., (2016). *Indian J. Phys., 90*(2), 155–162.

8. Behera, S., Das, P. R., Nayak, P., & Patri, S. K., (2017). *J. Elect. Mat., 46*, 1201.
9. Parida, B. N., & Das, P. R., (2014). *Journal of Alloys and Compounds, 585*, 234–239.
10. Biswal, L., Das, P. R., Behera, B., & Choudhury, R. N. P., (2012). *J. Electroceram, 29*, 204–210.
11. Maeder, M., Damjanovic, D., & Setter, N., (2004). *J. Electroceram., 13*, 385–392.
12. Pradhan, D. K., Behera, B., Das, P. R., & Das, P. R., (2012). *J. Mater. Sci Mater. Elec., 23*, 779.
13. Saparjya, S., Behera, S., Behera, B., & Das, P. R., (2017). *J. Mat. Sci; Mat. Elect., 28*, 3843.
14. Magasinski, A., Dixon, P., Hertzberg, B., Kvit, A., Ayala, J., & Yushin, G., (2010). *Nature Materials, 9*(4), 353–358.
15. Kityk, I. V., Makowska-Janusik, M., Fontana, M. D., Aillerie, M., & Fahmi, A., (2001). *J. Appl. Phys., 90,* 5542.
16. Dai, Z., & Akishige, Y., (2010). *J. Phys. D Appl. Phys., 43*, 4454031.
17. Singh, N., Agarwal, A., Sanghi, S., & Singh, P., (2011). *J. Magn. Magn. Mater., 323*, 486.
18. Jawahar, A., & Choudhary, R. N. P., (2008). *Mater. Lett., 62*, 911.
19. Behera, S., Parida, B. N., Nayak, P., & Das, P. R., (2013). *J. Mat. Sc, Mat. Electron, 24*, 1132.
20. Pilgrim, S. M., Sutherland, A. E., & Winzer, S. R., (1990). *J. Amer. Ceram. Soc., 73*, 3122–3125.
21. Macdonald, J. R., (1987). *Impedance Spectroscopy*. New York: Wiley.
22. Almond, D. P., & West, A. R., (1983). *Solid State Ion., 11*, 57.
23. Bechir, M. B., Karoui, K., Tabellout, M., Guidara, K., & Rhaiem, A. B., (2014). *J. Alloys Compd., 588*, 551.
24. Macedo, P. B., Moynihan, C. T., & Bose, R., (1972). *Phys. Chem. Glasses, 13*, 171.
25. SambasivaRao, K., Murali, K. P., Madhava, P. D., Lee, J. H., & Kim, J. S., (2008). *J. Alloys Comp., 464*, 497.
26. Mizaras, R., Takasighe, M., Banys, J., Kojima, S., Grigas, J., Hamazaki, S. I., & Brilingas, A., (1997). *J. Phys. Soc. Jpn., 66*, 2881.

CHAPTER 8

Advanced Materials for Electromagnetic Shielding

SUJI MARY ZACHARIAH,[1] ANANTHU PRASAD,[1] AVINASH R. PAI,[2] YVES GROHENS,[3] and SABU THOMAS[1]

[1]*International and Inter-University Center for Nanoscience and Nanotechnology (IIUCNN), Mahatma Gandhi University, Kottayam–686560, Kerala, India,*
E-mail: sujimaryzachariah@gmail.com (S. M. Zachariah)

[2]*School of Chemical Sciences, Mahatma Gandhi University, Kottayam–686560, Kerala, India*

[3]*Christian Huygens Research Center, Rue de Saint-Maudé, Lorient–56100, France*

ABSTRACT

The widespread use of electronic gadgets, wireless communication technologies, etc., has led to electromagnetic (EM) wave radiation creating a huge threat to biological systems, defense technologies, and commercial appliances. In regards to its mitigation, polymeric materials have gained huge attention from all over the globe. In this chapter, we aim to explore various materials like polymers, textiles, ceramic, and cement employed for controlling the EM radiations.

Advances in Diverse Applications of Polymer Composites: Synthesis, Application, and Characterization.
Suji Mary Zachariah, Yang Weimin, Maciej Jaroszewski, & Sabu Thomas (Eds.)

8.1 INTRODUCTION

There have been an explosive penetration of technological devices utilizing electromagnetic (EM) waves after the development of EM theory in the 19th century. These EM waves provide information transfer between distant terminals like satellites and space shuttles and are used for different purposes like broadcasting, wireless information transfer, imaging, and other related medical applications. On the other hand, they also create a new form of invisible and intangible pollution known as EM smog or electromagnetic interference (EMI). Interaction of EM waves from different sources results in interference or malfunctioning of equipment leading to financial/ data loss. Furthermore, they can also cause various health ailments like insomnia, nervousness, and headache. However, current research shows that long-term exposure to EM radiation can have even more serious effects on human health. These researches give evidence that a number of neuropsychiatric disorders, depression, attention deficit disorder, hyperactivity of children, and suicidal tendencies can be caused by EM radiation [1]. In relation to EM radiation, it is often discussed its impact on male fertility and sperm production, particularly in the case of putting a mobile phone in a pocket near the testicles. As per the study of Erogul et al. [2], it is proved that adverse effects of radiation on sperm really exist. Therefore, this has become a serious problem, and its mitigation could be achieved through the use of EMI shielding materials.

EMI shielding refers to blocking the EM waves with the help of a shield comprising of either conductive or magnetic-based materials [3–5]. This phenomenon has gained a greater importance today, owing to our complete dependence on electronics and also, due to the increased growth of RF radiating sources [3]. Depending on the need, there are different types of materials for shielding purposes like screens, curtains, and wall-papers used in hospitals, banks, airports, etc. In this chapter, we provide a detailed insight into various effects, mechanisms, and materials related to EMI shielding.

8.2 NEED FOR POLYMER-BASED SHIELDING MATERIALS

Several types of shielding materials are currently in use, and research that challenges the problem created by the EM radiation. Metals, due to their excellent conductivity towards heat and electricity, ability to absorb, reflect, and transmit stray radiations, have been a common choice for EMI shielding applications. However, poor flexibility, heaviness, cost, processing difficulty, corrosiveness, poor wear, scratch resistance, etc., have made scientists think of better options. Metals like Cu and Al were extensively used for shielding applications due to their excellent conductivity and mechanical properties. But for certain applications like aircraft, aerospace, or next-generation flexible electronic devices where weight is a criterion, metals are not preferred. Fillers like Iron-based nano-sized materials (e.g., ferrites, ferric oxide, magnetite, carbonyl ion, and Wustite), carbonaceous materials (e.g., graphite, graphene, graphene oxide (GO), reduced graphene oxide (rGO), carbon nanotubes (CNT), and carbon fibers (CF)), 2D metal carbides/ nitrides, etc., are very promising candidates when employed along with polymers to form blends, composites, inks, paints, etc., as they impart exceptionally high shielding value. Additionally, they also impart certain properties like low density, low fabrication cost, moldability, and high strength to weight ratio making polymer blends/ composites play a very prominent role in EMI shielding applications.

8.3 SOURCES

More than half of the world's population lives in an EM environment with a much higher EM field than existed 100 years ago [6]. Often magnetic and electric fields are considered to be the major sources of EMFs and are produced when a changing electric field produces a magnetic field or changing magnetic field produces an electric field. They have the capability to travel through empty space as well as through air and other substances.

In general, the sources of EMF in the occupational and residential environment can be classified as natural and man-made sources. The major natural source includes the Earth's magnetic field, thunderstorm, and lightning activity. Apart from natural sources, many man-made sources contribute to EM radiation. These include overhead high voltage power lines, dissemination from telecommunication devices like radar, radio, mobile phones, TV broadcasting, and equipment in hospitals for imaging, diagnostics, etc.

EM waves possess the feature of self-propagation in a vacuum or in any medium. They can be categorized based on the frequency as radio frequency waves (RF), microwaves (MW), visible light, ultraviolet waves (UV), X-rays, and gamma rays, as shown in Figure 8.1. Table 8.1 depicts different EM waves, frequency bands, and their application areas.

8.4 EFFECT OF EMI

In spite of the positive aspects of technological innovations that make life easier, it also involves many factors that impair the quality of life. For instance, inventions, and innovations in the communication sector have helped society in numerous ways. It has become pretty easy to get access to relevant information at anytime and anywhere, improved communication, etc. But its adverse effects were found to be countless, including health ailments, the lost art of conversation, and so on [7].

Numerous medical studies correlating EMF with serious health issues have been reported. Adverse pregnancy outcomes, including miscarriage, preterm delivery, altered gender ratio, and congenital anomalies have all been linked to maternal EMF exposure. International Journal of Cancer recently published a work on the relationship between childhood leukemia and magnetic fields in Japan. By assessing the level of the magnetic field in children's bedrooms, scientists confirmed that high EMF exposure was the major reason for childhood leukemia [8]. Amongst other symptoms, reports regarding sleep disturbances due to EMF have been noted in anecdotal reports. Although the exact neurobiological mechanisms related to sleep are not yet known in detail, the regular sequences of waking and sleeping states are essential for correct information processing of the brain, metabolic homeostasis, and intact immune function. Moreover, sleep appears to be an appropriate physiological system to be studied with the aim of elucidating the interaction between high-frequency EMF and the human organism as sleep is a well-defined biological condition, reacting very sensitively to external influences [8, 9]. Studies on operators of dielectric welders and HF (high frequency) presses revealed that exposure to high-intensity EMFs can cause severe eye irritation, numbness on fingertips, lower heart rate, etc. [10, 11].

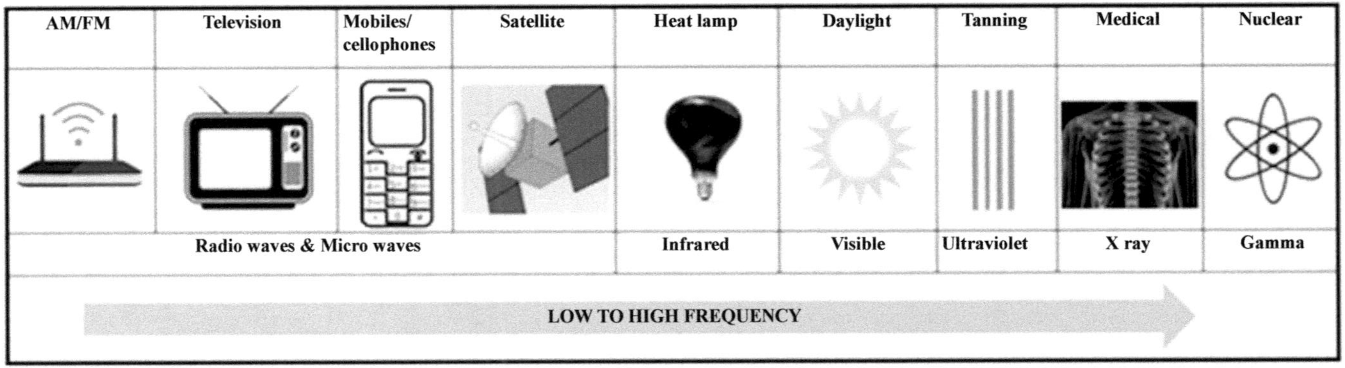

FIGURE 8.1 Different EM waves are based on the frequency and their sources.

TABLE 8.1 Different EM Waves, Frequency Bands, and Their Application Areas

EM Wave	Frequency (Hz)	Band	Application Areas
Radiofrequency	3–30 kHz	Very low frequency	Beacon
	30–300 kHz	Low frequency	Marine communications
	300 kHz–3 MHz	Medium frequency	Radio broadcasting
	3–30 MHz	High frequency	Radio frequency identification
	30–300 MHz	Very high frequency	Radio broadcasting, television
Microwave	300 MHz–1 GHz	Ultra-high frequency	Television, microwave oven, mobile phones
	1–2 GHz	L band	GPS, RADAR, wireless LAN, mobile phones
	2–4 GHz	S-band	Bluetooth, cordless phones, television, mobile phones
	4–8.2 GHz	C and J band	Satellite communication, cordless phone, WIFI
	8.2–12.4 GHz	X band	Satellite communication, weather monitoring, air traffic control, defense tracking
	12.4–18 GHz	Ku band	Satellite communication
	18–27 GHz	K band	Satellite communication
	27–40 GHz	Ka-band	Satellite communication
	40–75 GHz	V band	Military and research
	75–110 GHz	W band	Military and research

8.5 FACTORS AFFECTING EMI SHIELDING EFFECTIVENESS (SE)

Ideal material for mitigating the effects of EM waves should possess certain characteristics like impedance matching, dielectric, and magnetic loss property, dielectric polarizability, permeability, electrical conductivity, frequency, etc. [12, 13]. Each of these is discussed briefly below.

An ideal EMI shielding material, first of all, requires impedance matching characteristics between free space (air) and the surface of shielding material, thereby hindering reflection. In other words, an impedance of free space and impedance of material should match thereby letting EM waves enter into the absorbing material to a greater extent. In simple terms impedance matching characteristic is that the incident EM wave can enter into the material or absorbed into the materials to a greater extent. The condition for complete impedance matching is $^{1}/\varepsilon^{1}$ to be equal to 1.

$$\varepsilon^* = \varepsilon^{I} - j\varepsilon^{II} \tag{1}$$

where; ε^{I} and ε^{II} are the real and imaginary part of dielectric constant. The real part represents the energy storage or charge capacity while the imaginary part is known as dielectric loss factor which represents the energy loss.

$$\mu^* = \mu^{I} - j\mu^{II} \tag{2}$$

where; μ^{I} and μ^{II} are the real and imaginary part of permeability. Here both μ^{I} and ε^{I} varies significantly with the frequency.

The ratio of an imaginary part to the real part of permittivity is known as electric loss tangent.

$$\tan\delta e = \varepsilon''/\varepsilon^{I} \tag{3}$$

Greater is the electric loss tangent; greater is the attenuation as wave travel through the medium. The ratio of the imaginary part to the real part of permeability is known as magnetic loss tangent.

$$\tan\delta m = \mu''/\mu^{I} \tag{4}$$

Greater permeability more will be the tendency to absorb the EM waves. Therefore, permittivity, and permeability are crucial parameters to design an effective EMI shielding material, as explained in the previous section.

Polymer composites with large dielectric constants are very essential for providing dielectric loss due to the conductivity mismatch between conductive filler and insulating polymer matrix. A difference between electrical conductivities of adjacent materials results in polarization and charge accumulation at their interfaces. And with this synergistic effect, high 'k' value can be obtained. The filler type and its orientation can also increase the polarization ability of the polymeric system [14].

Similarly, the thickness is also a factor in controlling EMI SE. The absorption loss increases with the increasing thickness of the shield, whereas the reflection loss is independent of the shield thickness. Generally, high values of SE can be obtained with increasing material thickness, but there is always some limitation in this regard when taking the cost and density requirements into account [6].

Frequency of incident EM field is another factor affecting SE. The reflection loss decreases with increasing frequency, while the absorption loss increase with increasing frequency. The decrease of reflection loss with increasing frequency occurs due to the increase of shield impudence with frequency, whereas absorption loss increases with frequency because of the decrease of skin depth with frequency. The SE in case of reflection and absorption is given by Eqns. (5) and (6).

$$SE_R = 39.5 + 10\, Log\frac{\sigma}{2\pi\mu} \tag{5}$$

$$SE_A = 8.7d\sqrt{\pi f \mu\sigma} \tag{6}$$

where; σ is electrical conductivity; μ is permeability; and 'f' is frequency. These equations can clearly illustrate the variation of reflection loss and absorption loss with frequency [15].

8.6 MECHANISM OF SHIELDING

EMI shielding occurs by mainly three mechanisms, namely reflection (a primary mechanism), absorption (a secondary mechanism), and multiple reflections. For the primary mechanism to work the shielding material must contain conducting particles or mobile charge carriers (electrons and holes) so that they will interact with the incident radiation. In the case of a secondary shielding mechanism which is the most preferable mechanism to arrest harmful radiations, the material must contain materials with high dielectric constant or high permeability. For such purpose, materials with high dielectric constant like ZnO (zinc oxide), SiO_2 (silicon dioxide), TiO_2 (titanium dioxide), and $BaTiO_2$ (barium titanate) are used. Materials with high magnetic permeability used include Ni, Co, carbonyl iron, Fe_3O_4, Mu-metal, and super permalloy. Other than the primary and secondary mechanisms, another mechanism of shielding is multiple reflections which correspond to the reflection from multiple surfaces or interfaces in the shield. This particular mechanism requires the presence of large interfacial area and porous structure, for instance, a conducting composite material consisting of filler, foamed composite, and honeycomb structures [6, 16].

The EM shielding performance is described by the term EMI shielding effectiveness (SE_T) which measures how well a material impedes the EM energy of a certain frequency when passing through it [12]. It is usually quantified in terms of the logarithm of the incident power over the transmitted power (or electric or magnetic field). It can also be expressed in terms of contribution from reflection, absorption, and, multiple internal reflections (Eq 8) as per Schelkunoffs theory.

$$SE_T = 10log\left({}^{P_I}\!\Big/_{P_T}\right) = 20log\left({}^{E_I}\!\Big/_{E_T}\right) = 20log\left({}^{H_I}\!\Big/_{H_T}\right) \tag{7}$$

$$SE_T = SE_R + SE_A + SE_M \tag{8}$$

where; P, E, and H indicate power, electric, and magnetic field while; I, R, and T represent incident, reflected, and transmitted components.

The reflection loss (SE_R) value is associated with the relative mismatch between the incident EM wave and surface impedance of the shielding material. This is represented by the Eqn. (9).

$$SE_R = -10\log_{10}\frac{\sigma_t}{16f\varepsilon\mu_r} \tag{9}$$

where; σ_t is the total conductivity (S/cm); 'f' is the frequency; 'ε' is the electric permeability; and μ_r is the relative magnetic permeability. This equation implies that SE_R is a function of the ratio of σ_t and μ_r. Additionally, this equation also implies that SE_R decreases with increasing 'f' value with fixed σ_t and μ_r for given shielding material.

The absorption loss (SE_A) value is a physical characteristic of the shielding material and is independent of the type of source field. Absorption loss or decay occurs due to ohmic loss and heating of material. This is represented by Eqn. (10).

$$SE_A = 3.34t\sqrt{f\sigma_r\mu_r} \tag{10}$$

In the case of multiple reflection mechanism, the reflected wave gets bounced and re-reflected continuously between the boundaries. This is represented by Eqn. (5) that shows a close relation between SE_M and SE_A.

$$SE_M = 20\log_{10}\left(1-e^{\frac{2t}{\delta}}\right) = 20\log\left(1-10^{\frac{SE_A}{10}}\right) \tag{11}$$

SE_M plays an important role in porous structures, multilayer structures and in certain other morphological and geometric structures. This value could be neglected for thicker shields as the amplitude of the absorbed EM wave becomes negligible as it reaches the next boundary, i.e., when SE_A is high (≥ 10 dB). In other words, SE_M is significant only for thin metals and when used at low frequencies (kHz frequency range).

Shielding values > 10 dB (decibels) offer a very limited range of shielding. For values between 10 dB and 30 dB, a minimum effective range of shielding is obtained, which is suitable for commercial applications, and if the values are > 30 dB, it could be very efficiently used for both industrial and commercial applications.

8.7 EXPERIMENTAL TECHNIQUE TO MEASURE SE

Network analyzer instruments that operate on the principle of the waveguide technique are used for evaluating EMI SE. Two types of network analyzers are used, namely: scalar network analyzer (SNA) and vector network analyzer (VNA). SNA works to determine the amplitude of signals, but cannot be used for determining complex permeability and permittivity. VNA is the most preferred instrument which is used to detect the signal magnitude response as well as phases of various signals. Network analyzers basically consist of a signal generator, receiver, test set, and display. Most of the VNA consists of two test ports used for measuring four S-parameters, namely S_{11}, S_{21}, S_{12}, S_{22} where they represent the reflection and transmission coefficient value of the sample material. In other words, one could measure the S-parameters and can calculate the reflectance and transmittance power coefficient using the Eqns. (12) and (13).

$$R = |S_{11}|^2 \quad (12)$$

$$T = |S_{21}|^2 \quad (13)$$

8.8 MATERIALS USED

As known to everyone EM waves harmfully affect both the device performance and human beings, but a reduction in its usage is not practical. A better way of its mitigation is by reducing the penetration of EM wave produced by providing a shield or blocks the EM waves from the desired surface. Main materials used for shielding applications used include metals, polymers, textiles, cement, and ceramics (Figure 8.2). Metals in the form of thin sheets were commonly used for EMI shielding but due to cost, weight, corrosiveness, etc., makes them an undesired choice. In this scenario, polymer composites have attracted a great deal of academic and industrial interest due to its cost-effectiveness, ease of processability, and many other practical applications [6]. But their inability to form thin films as compared to metal films is a major drawback. Recently conductive inks/ paints also have gained huge interest as they pose good features like good processability, flexibility, and tiny thickness. As the majority of the polymers are insulators in nature, they need to be modified using certain filler materials for EM shielding application. They include carbonaceous materials like CB, graphite, graphene, GO, CNT, CNF, dielectric materials like TiO_2, $BaTiO_3$, magnetic materials like Fe_2O_3, $CoFe_2O_4$, etc.

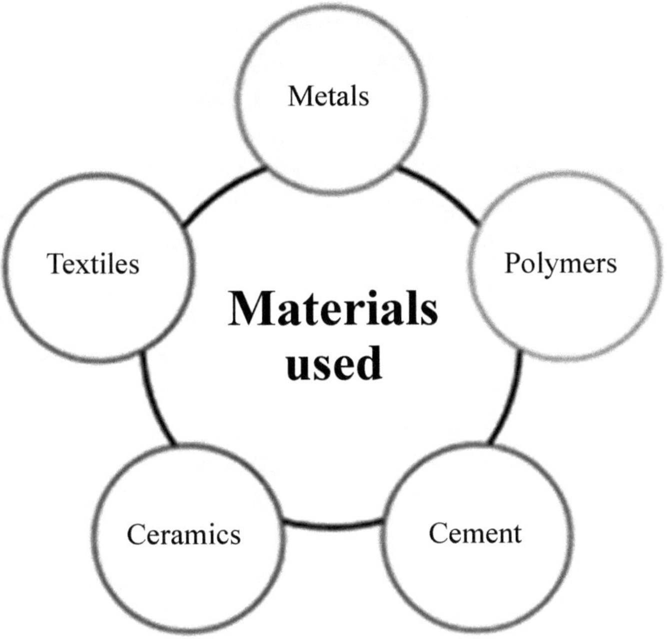

FIGURE 8.2 Different materials are used for EM shielding applications.

A point to be noted is that among the filler materials mentioned above, magnetic fillers have an inability to get uniformly dispersed in polymer matrix. But could be solved by combining them with carbonaceous fillers, which facilitate the dispersion in polymer matrices by increasing the viscosity of polymer melts. The following section describes in detail various polymeric and non-polymeric materials developed for EMI shielding applications.

8.8.1 INSULATING POLYMER-BASED EMI SHIELDING MATERIALS

Most of the polymers are considered to be inherent insulators, but could be made conducting by incorporating with conducting fillers as mentioned in Section 8.2. These are discussed in detail below.

8.8.1.1 CARBON-BASED FILLERS SHIELDING MATERIALS

CB which is a pure elemental form of carbon is a prominently used chemical in the tire industry as a reinforcing agent. Apart from imparting strength and durability, they also impart conducting nature to the host matrix. Based on these features, Zhou et al. [17] fabricated wood/polyethylene composites (WPCs) by addition of a small amount of nanoscale conductive carbon black (CCB), leading to honeycomb-distributed WPC ($WPC_{H\text{-}CCB}$). Additionally, they also prepared uniformly-distributed WPC ($WPC_{U\text{-}CCB}$) by melt-blending process and compared their conductivity, EM interference shielding (EMI SE), and flame-retardancy. Conductivity studies using a four-probe tester created a honeycomb network at < 3% of CCB loading in $WPC_{H\text{-}CCB}$ resulting in higher conductivity values than $WPC_{U\text{-}CCB}$. The SE value for the $WPC_{H\text{-}CCB}$ was found to be about 20 dB at 3 wt.% of CCB loading while for the $WPC_{U\text{-}CCB}$ it was <5 dB in the X band. In short, the shielding ability of $WPC_{H\text{-}CCB}$ was found to be considerably greater than $WPC_{U\text{-}CCB}$ even at highest CCB content. Combustion studies were conducted using cone calorimeter apparatus to study the flame retardancy with a heat flux of 50 kW. m^{-2}. As inferred from Figure 8.3, WPC without CCB coating (WPC_{CTRL}) produced considerable smoke during combustion, whereas smoke suppression was better for $WPC_{H\text{-}CCB}$ due to the sufficient coverage of CCB that shields the material from heat and O_2. Additionally, $WPC_{H\text{-}CCB}$ had the lowest SPR (smoke production ratio) and TSP (total smoke production), and higher mass retention implying the excellent fire retardancy feature of $WPC_{H\text{-}CCB}$.

Yuan et al. [18] recently developed a facile strategy to convert waste epoxy based composites for EM shielding and strain sensing application. They proposed an interface engineering strategy and recycled the waste epoxy while enhancing the mechanical and electric properties at low filler content. Initially, they infused a catalyst and later carbon nanotube structures (CNS) into the recycled epoxy system. Investigations reveal that this CNS/epoxy composite exhibit excellent tensile and electrical properties. For instance, comparing with neat epoxy, 0.25 wt.% loaded composite showed an enhancement in modulus and strength by 1,757% and 252%, respectively, which is attributed to the uniform dispersion of CNS at low filler content. This aids the epoxy chains to penetrate throughout the interconnected filler network thereby providing exceptional mechanical reinforcement to the composite. But increasing the filler content develops agglomeration resulting in a minor reinforcement. Additionally, the electrical percolation (ϕ_c) values obtained were very less by 0.01 wt.% whereas, for composites obtained by conventional methods was between 0.05 wt.% and 0.1 wt.%. The electrical conductivity studies showed that the values were superior to that of conventionally prepared CNS/epoxy composite. Besides the low ϕ_c value and superior conductivity values, these composites could be processed again by adding a small amount of catalyst. Furthermore, these features enable the composite to be also used as a strain sensor.

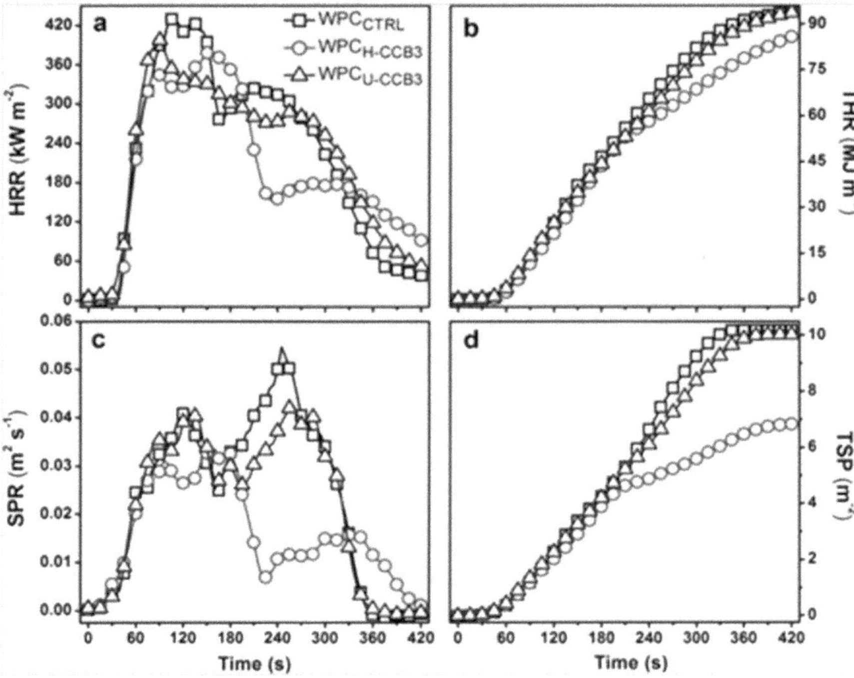

FIGURE 8.3 (a) Heat release rate (HRR); (b) total heat release (THR); (c) smoke production ratio (SPR); and (d) total smoke production (TSP) of WPC_{CTRL}, WPC $_{U\text{-}CCB3}$, and WPC $_{H\text{-}CCB3}$.

Source: Reprinted with permission from Ref. [17]. © 2020 Elsevier.

Joshi et al. [19] developed a thin, lightweight polyvinyl alcohol/graphene nanoribbons (PVA/GNR) composite film with excellent shielding value at low filler concentration. A SE value of 60 dB was obtained at 0.025 wt.% of filler loading with a sample thickness of 0.3 mm. Additionally, the composite film showed 14 orders of increase in conductivity with filler concentration of 0.0075 wt.%. AFM images (Figure 8.4) reveal that a particular alignment of GNR in the polymer matrix provided a path for electron transfer and therefore helped in enhancing electrical conductivity.

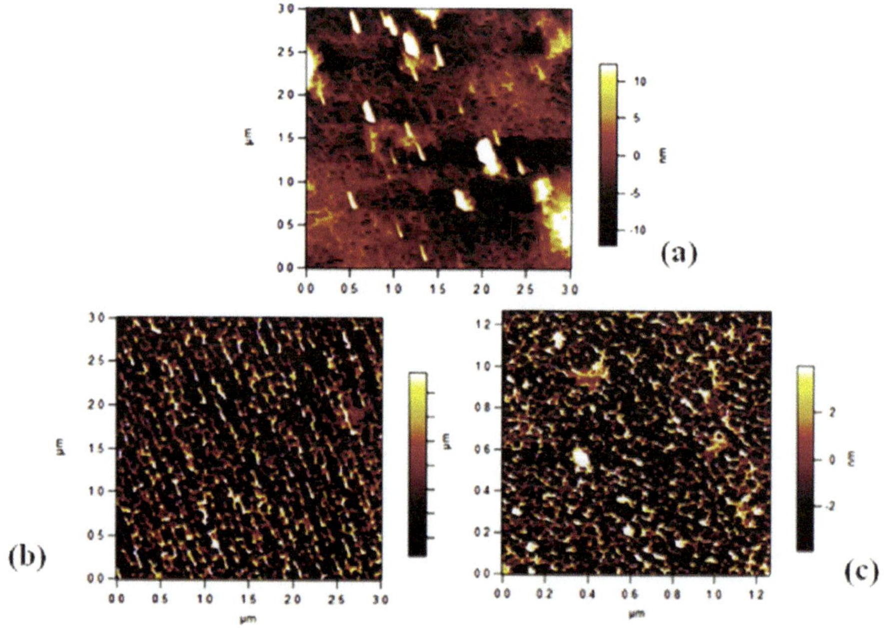

FIGURE 8.4 (a) AFM image of neat PVA; (b) and (c) AFM images of composite in different areas.

Source: Reprinted with permission from Ref. [19]. © 2013 IOP Publishing Ltd.

8.8.1.2 MAGNETIC FILLER-BASED SHIELDING MATERIALS

Recovering metal components from various sources (particularly from wastes) and their utilization in the production of new composite material where they bring benefits of enhancing the final properties and reducing the environmental burdens is a versatile concept. For instance, there were numerous studies conducted on recovering metals from discarded printed circuit boards (PCB) and other items. La Rosa et al. [20] used metal fillers recycled from PCB along with HDPE (high-density polyethylene) recovered from municipal solid wastes (MSW) as sandwich panels for structural applications (refer to Figure 8.5). Two different types of panels were fabricated via injection molding after extrusion with a final thickness of 2.5 mm. The shield type 1 consists of three layers, namely glass wool, adhesive, and Al foil, while shield type 2 consists of a PS sheet, adhesive, and Al foil. Glass wool and PS sheet function as thermal insulating materials, while Al foil enhances the SE values. The SE of the samples were studied at a frequency of 10 GHz and the optimum attenuation value for the shield type 1 and 2 obtained was 48.3 and 42.7 dB, respectively.

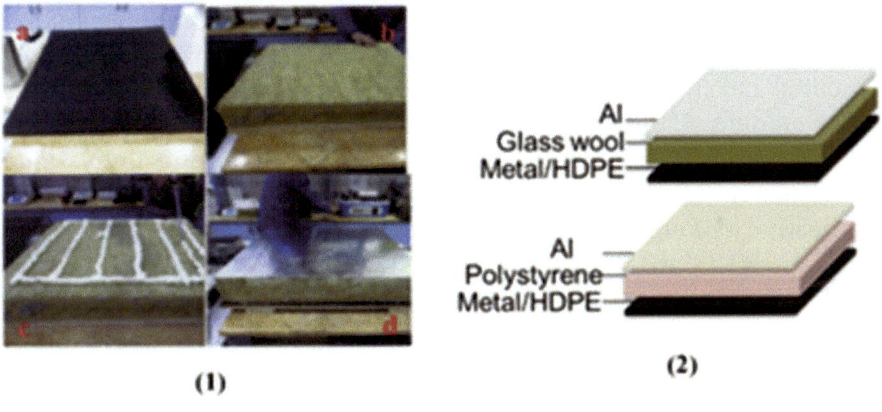

FIGURE 8.5 (1) Schematic representation of the shield Type 1: (a) CPC sheet; (b) glass wool; (c) adhesive; (d) aluminum foil. (2) Snapshots from the fabrication of the EM shield Type 1.

Source: Reprinted from Ref. [20]. https://creativecommons.org/licenses/by/4.0/

Aerogels are synthetic, lightweight, porous materials with high internal surface area. Addition of magnetic and conductive particles could help in fabricating lightweight EM shielding material with high shielding value. Keeping these advantages in mind, Chen et al. [21] developed cellulose/rGO/Fe_3O_4 aerogel using an environmentally friendly, simple, efficient approach. Analysis reveal that for 8 wt.% rGO and 15 wt.% Fe_3O_4 content with 0.5 mm sample thickness, SE of 32.4–40.1 dB in the X band frequency range was obtained. Upon increasing the rGO content and sample thickness, the shielding performance-enhanced greatly. Additionally, multi-reflection at the cell walls and impedance matching due to the porous structure, rGO sheets, and Fe_3O_4 nanoparticles has tuned this aerogel into a highly efficient EM shielding material. The entered microwaves are scattered and reflected multiple times at the rGO sheets, leading to lengthening of the propagation paths of the microwaves in the aerogel, which further improved the absorption capacity of the cellulose aerogel.

8.8.1.3 DIELECTRIC FILLER-BASED SHIELDING MATERIALS

Singh et al. [22] prepared composites based on different metal oxides like Fe_2O_3, SiO_2, ZnO, ZrO_2, and TiO_2 in PVA matrix by solvent casting method and studied the SE in both C band and X band frequency range. The amount of metal oxides in composites, ranged from 0.1 to 10 wt.%. As the concentration increased the minimum reflectivity values also increased gradually. At 10 wt.% filler loading, the minimum reflectivity of composites was found to be −38.85 dB at 10.4 GHz for Fe_2O_3, −41.90 dB at 10.4 GHz for SiO_2, −33.65 dB at 10.4 GHz for ZnO, −24.90 dB at 11.0 GHz for ZrO_2, and −32.90 dB at 9.76 GHz for TiO_2. Furthermore, studies revealed that the losses in PVA composites based on SiO_2, ZnO, ZrO_2, and TiO_2 were dielectric in nature, whereas PVA/Fe_2O_3 composite exhibited both, dielectric, and magnetic losses. The results also demonstrated that composites based on Fe_2O_3 and SiO_2 exhibited good shielding property, mechanical strength and high thermal stability and were reported to be low-cost, efficient EMI shielding materials.

8.8.1.4 MIXED FILLER-BASED SHIELDING MATERIALS

Kriti et al. [23] reported the formation of $BaTiO_3$/$CoFe_2O_4$ (BT-CFO) filled low-density-polyethylene (LDPE) composite at different filler percentages (4 to 20 wt.%) via melt mixing and studied their dielectric and EM shielding behaviors. The dielectric behavior of BT-CFO/LDPE composites exhibited a change in dielectric constant with an increase in temperature and frequency. The EM shielding measurements in the X-band exhibited a maximum value of 17.9 dB for 8 wt.% filler loading in comparison with pure LDPE with 11.6 dB shielding value.

8.8.2 INTRINSICALLY CONDUCTING POLYMER (ICPS)-BASED EMI SHIELDING MATERIALS

Intrinsically conducting polymers (ICPs) are extensively employed in the electronics industry due to their electrical conductivity and flexibility properties. But their synthesis, cost, and availability have been a limiting factor for its optimum level of commercialization and limit their use in the industrial sector. Fortunately, their excellent electrical conductivity in doped states, make them very versatile for EM shielding applications. Delocalization of π electrons in their conjugated structure has imparted this unique electronic property, making them suitable for shielding application. But extensive delocalization could hinder their processability. Problems like swelling, cracking, contraction, etc., could also occur, thereby affecting the mechanical and electronic properties. These problems are surpassed by adding materials like metal oxides, magnetic materials, metal nanoparticles, and carbon nanostructures. Addition of these could also enhance thermal stability, dielectric, and magnetic property without compromising their electrical conductivity. Hence ICPs like polyaniline (PANI), polypyrrole (PPy), polythiophene (PT), and their composites are suitable alternatives for EM shielding applications. Among these, PANI has attracted a major attention due to its simple, eco-friendly synthesis, and facile doping technique to attain adequate electrical conductivity.

8.8.2.1 CARBON-BASED SHIELDING MATERIALS

Most of the foamed metal-based composites pose better SE values but suffer the problem of high density. Recently Wu et al. [24] developed a carbon-based composite that overcome the density problem with superior SE value. The composite was fabricated by drop coating poly(3,4-ethylene dioxythiophene) (PEDOT) and polystyrene sulfonate (PSS) solution on dodecylbenzene sulfonic (DBSA) acid-functionalized graphene foam (GF). An ultralow density value of 0.0182 g/cm^3 and 98.8% porosity were obtained with 58% weight fraction of GF. Additionally, the composite gave superior electrical conductivity values of 43.2 S/cm and SE value of 90 dB, which is attributed to the capacitor-like behavior induced by polarization and resistor-like features originating from the conductive composite networks.

There has been a growing attention in recent years in the area of nano-cellulosic substrates for the preparation of PPy-based flexible conducting composites. Some of these include carbon nanofibers (CNF) and cellulose nanocrystals (CNC). CNF is more promising due to its intrinsic flexibility, biodegradability, lower cost of production, ability to interact with the PPy through hydrogen bonding thereby improving processability of PPy. But CNC is relatively expensive, rigid, and form brittle films in comparison with CNF. Furthermore, PPy usage in EM shielding is very prominent due to its high electrical conductivity, good thermal and environmental stability, low toxicity, and easy synthesis in aqueous solvents [25, 26].

Recently Parit et al. [27] developed a cost-effective composite film using PPy and CNF by in-situ polymerization of PPy onto cellulose nanopaper (CNP) and PVA coated cellulose nanopaper (PVA-CNP). The newly developed composite showed enhancement in the conductivity, mechanical properties, and SE value when compared with commonly used *in-situ* nanofiber (ISF) approach, where PPy was coated on nanofibers prior to film preparation. PVA coating imparted a smooth surface for hydrogen-bonding interactions making a continuous and compact PPy network that is responsible for enhanced conductivity (8.4 S/cm) in PPy/PVA-CNP than in PPy/CNP. As SE value is directly related to conductivity, SE value is also higher for PPy/PVA-CNP composites. The ISF sample exhibited a SE value of 10.6 dB, while the PPy/CNP and PPy/PVA-CNP composites had a higher SE value of –21 dB and –23 dB, respectively. Similarly, tensile strength was also higher for PPy/PVA-CNP than for PPy/CNP which is due to the reduced porosity because of the impregnation of PVA into CNP.

8.8.2.2 MAGNETIC FILLER-BASED SHIELDING MATERIALS

Agrawal et al. [28] synthesized PPy from its monomer using a facile and eco-friendly strategy. Easy formation of PPy using $FeCl_3.6H_2O$ as the oxidant in water makes the synthesis route eco-friendly. The reactant mixtures were mixed at different reaction durations for 0.3, 1, 2, 4, 6, 8, 10, and 24 hr. at 5°C and were denoted as PPy-1, PPy2, PPy-3, PPy-4, PPy-5, PPy-6, PPy-7, and PPy-C. The conductivity values and SE of PPy-6 samples were found to be higher, which was 2S/cm and 30 dB, respectively. Investigation showed that presence of Fe_2O_3 at very low concentration helped in achieving a better SE value in the X-band region. It was also noted that the conductivity of PPy varied with reaction time as seen in Figure 8.6. Furthermore, it is also noted that conductivity increases with temperature and vice versa as lattice vibration reduces at lower temperature.

Red mud (RM) or bauxite tailings are a toxic by-product obtained during the production of alumina. It mainly consists of metallic oxides like Fe_2O_3, Al_2O_3, SiO_2 and TiO_2. These oxides help in absorption of EM waves and keeping this in view, Pande et al. [29] developed a composite for EM shielding application by in-situ chemical oxidative polymerization of aniline. The PANI/RM composites have shown SE of 33–41 dB in the X-band frequency range for 3 mm sample thickness. A core-shell type of morphology was developed as shown in the TEM micrographs (Figure 8.7), which reveal that the RM particles have polymerized along with aniline, resulting in this morphological structure and finally in better EM wave absorption.

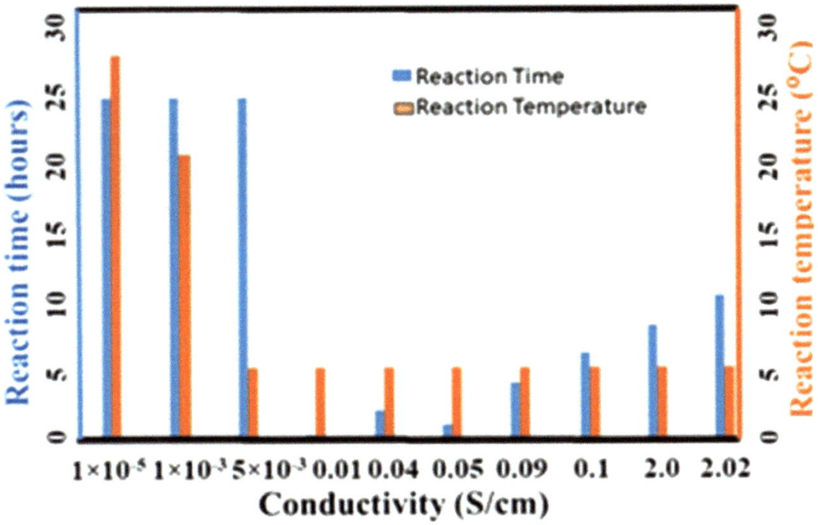

FIGURE 8.6 Room temperature DC conductivity values of polypyrrole samples vs. reaction time and reaction temperature.

Source: Reprinted with permission from Ref. [28]. © 2020 Wiley.

8.8.2.3 DIELECTRIC FILLER-BASED SHIELDING MATERIALS

Among various fillers used in shielding application $BaTiO_3$ has a prominent role due to their high permittivity, large attenuation constant, etc. But previous works based on conducting polymer-$BaTiO_3$ composites fabricated by ex-situ mixing process resulted in poor shielding performance due to non-uniform dispersion of $BaTiO_3$ in the polymer matrix. Therefore, Saini et al. [30] developed an in-situ mixing strategy to develop PANI/$BaTiO_3$ composites with good electric and dielectric properties that collectively contribute to better shielding performance by in-situ polymerization of aniline monomer in the presence of dodecyl benzene sulfonic acid dispersed in $BaTiO_3$ nanoparticle. SE analysis in the Ku band reveals that the attenuation level is very high for this composite, i.e., a total SE of 71.5 dB. This value represents EM attenuation > 99.9999% which means that < 0.00001% of the incident radiations has transmitted. This is attributed to the optimized electrical and dielectric attributes of the $BaTiO_3$ particles.

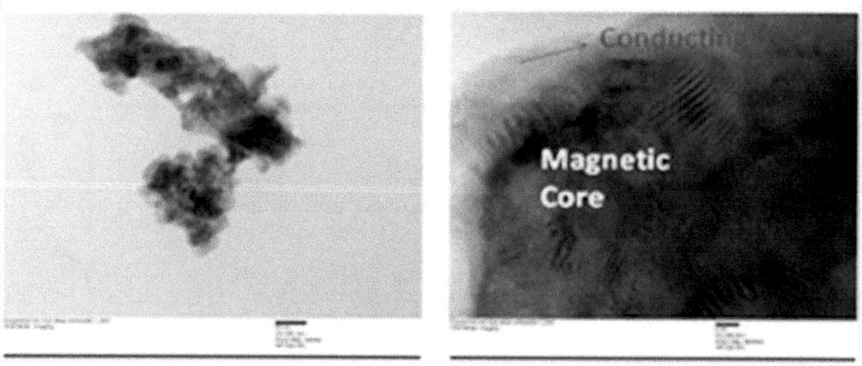

FIGURE 8.7 TEM images of RM/PANI composites.

Source: Reprinted with permission from Ref. [29]. © 2020 Elsevier.

8.8.2.4 MIXED FILLER-BASED SHIELDING MATERIALS

Li et al. [31] fabricated polyaniline/strontium ferrite/multiwalled carbon nanotubes (PANI/SrM/MWCNT) composites using an in situ polymerization approach. The electrical conductivity of the composites was seen to increase with MWCNT loading. For instance, for 0, 1, and 2 g of

MWCNT the conductivity values were 2.4853, 6.635, and 7.2196 S/cm, respectively. The EM properties of the composite were also excellent in 2–18 GHz frequency range. This is attributed to the better dielectric loss in the range of 2–9 GHz, and magnetic loss in 9–18 GHz frequency range.

Zeng et al. [32] fabricated a foldable paper using leather solid waste (LSW), polyvinyl alcohol (PVA), and silver (Ag) using a facile, sustainable electroless plating (ELP) technique. OH, groups in PVA cross-links with the active groups in debundled leather fibers (LFs) to generate LSW/PVA substrate and later coated with Ag layer to form LSW/PVA/Ag paper. The Ag layer imparted good hydrophobicity, reflection effect, tensile strength, and thermal stability to the paper. In addition to the reflection effect of metallic Ag on EM wave, the hierarchical structure of collagen fibers imparted a superior SE value of 55–90 dB in X band frequency range by an absorption-dominant mechanism. Furthermore, a multilayer LSW/PVA/Ag paper was also prepared with superior SE of 111.3 dB attributed to the construction of multiple reflection–absorption interfaces.

8.8.3 CERAMIC-BASED EMI SHIELDING MATERIALS

Ceramics are non-metallic inorganic materials that are brittle, hard, having excellent mechanical properties, wear resistance, chemical stability, satisfactory corrosion, and oxidation resistance properties. Additionally, they also possess the ability to work efficiently at higher temperatures. They are the least used materials for shielding applications due to their low conductivity nature. Similar to magnetic metals, there exist ceramics with magnetic property. However, they are not conductive like metals but are corrosion resistant. However, a synergistic use of ferrites with conducting particles like MWCNTs, r-GO, conducting polymers will be effective. Furthermore, there is a new class of ceramic called MXenes having features like excellent thermal stability, oxidation resistance, electrical, and thermal conductivity. Interestingly, most of the reported SE values for this class of ceramics are above 50 dB making them versatile materials for EMI shielding application.

8.8.3.1 CARBON-BASED SHIELDING MATERIALS

Carbon/ceramic-based composites are promising candidates as EM shielding materials to be used in very harsh environments. Keeping this view in mind Pan et al. [33] prepared two kinds of composite ceramics, one reinforced with carbon nanowires (CNWs) and other reinforced with nanowires-nanotubes (CNWs-CNTs). Results indicate that for the composites with 5.15 wt.% CNW loading, the SE_T value and SE_A reached 25.0 dB and 21.3 dB, respectively inferring that 99.7% incident signal can be blocked. For the CNW-CNT/Si_3N_4 composite with 3.91 wt.% loading the SE_T value was 25.4 dB.

By contrast, CNWs/Si_3N_4 showed better EM attenuation capability with stronger absorption coefficient, mainly due to its complicated networks, higher defect density and unique microstructure.

Barani et al. [34] fabricated a multifunctional graphene composite (epoxy/graphene/few-layer graphene) for packaging application of microwave components with excellent shielding value and heat conduction features. The composite showed excellent SE_T of $\approx$ 65 dB for sample thickness of 1 mm and 19.5 vol% filler loading in the X-band frequency range. At higher filler loading the composites also showed thermal conductivity of 11.2 ± 0.9 Wm^{-1} K^{-1}, which is comparable to the ceramics and the value is about 41 times larger than that of the pristine epoxy. Interestingly, the shielding efficiency improved as the temperature increased while the thermal conductivity remained constant. The enhancement in the shielding efficiency is attributed to the two electrical conduction mechanisms, namely, electronic band-type conduction inside the fillers, and hopping conduction between the fillers.

8.8.3.2 MAGNETIC FILLER-BASED SHIELDING MATERIALS

As mentioned previously, similar to magnetic metals, there exist magnetic ceramics like nickel ferries ($NiFe_2O_4$) that are highly efficient in shielding by absorption mechanism. They are non-corrosive but are not conductive. Recently, Yadav et al. [35] synthesized $NiFe_2O_4$ nanoparticles by dextrin mediated sol-gel combustion method followed by annealing at 600, 800, and 1,000°C. The nanocomposites consisted of $NiFe_2O_4$, rGO as the filler in the PP matrix (NFD-rGO-PP composite). The average crystallite size of nanoparticles was 20.6, 34.5, and 68.6 nm at annealing temperatures of 600°C, 800°C, and 1,000°C, respectively. Results revealed that the particle size played a prominent role in the EM properties and mechanical properties of nanocomposites. For instance, the composite exhibited SE_T of 45.56, 36.43, and 35.71 dB for particle sizes 20.6, 34.5, and 68.6 nm, respectively at 2 mm thickness in the X band frequency range, thus inferring that smaller-sized $NiFe_2O_4$ nanoparticles exhibits excellent EM shielding characteristics. Dependence of $NiFe_2O_4$ nanoparticles on mechanical properties is revealed from the stress-strain plot of the composite. The values of tensile strength and elongation at break are 1.89, 2.21, 2.36 MPa and 24.01, 77.38, and 259.34% for annealing temperatures of 600, 800, and 1,000°C. In addition, the Youngs modulus values decreased with increase in annealing temperature and are 33.34, 28.19, and 24.55 MPa, respectively.

8.8.3.3 MIXED FILLER-BASED SHIELDING MATERIALS

Qing et al. [36] fabricated $NiFe_2O_4/BaTiO_3$ ceramic for high-temperature shielding applications. $NiFe_2O_4$ nanoparticles were prepared by co-precipitation method and later mixed with $BaTiO_3$, then pressed by uniaxial pressing and finally sintered to form $NiFe_2O_4$ (30 wt.%)/ $BaTiO_3$ (70 wt.%) ceramics. For evaluating the influence of $NiFe_2O_4$ size on the EM properties $NiFe_2O_4/BaTiO_3$ ceramics with micron-sized particles were prepared by standard solid-state reaction method. The results showed that $NiFe_2O_4$ nanoparticles filled $BaTiO_3$ ceramic showed high complex permittivity and SE_T >34 dB in the X band frequency range than the corresponding micro-sized ones. Figure 8.8 shows that absorption plays a prominent role in the SE_T rather than reflection which is due to the enhanced interfaces polarization, and higher dielectric constant.

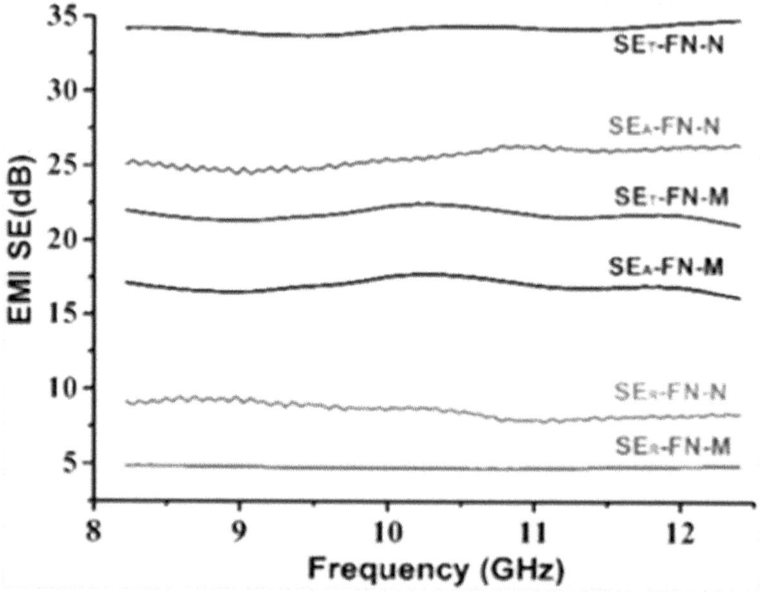

FIGURE 8.8 The effect of $NiFe_2O_4$ particle size on the EMI SE, SE_R, and SE_A of the $NiFe_2O_4/BaTiO_3$ ceramics with 1.2 mm thickness in the X-band.

Source: Reprinted with permission from Ref. [36]. © 2018 Elsevier.

Liu et al. [37] introduced a novel strategy to synthesize RGO/CNTs-SiCN ceramic nanocomposites with excellent shielding and dielectric value. A single-source precursor (SSP) was synthesized by chemical modification of poly(methylvinyl) silazane with chemically bonded GO and CNT and later warm-pressed, pyrolyzed, and thermally reduced to form RGO/CNTs-SiCN composite. The common feature of this novel system is that the dielectric properties of the composite prepared through SSP technique exhibits higher value than those prepared via physical blending. For instance, tangent loss vs. frequency plot (Figure 8.9) infers that sample S4 (SSP-derived RGO/CNTs-SiCN) has higher dielectric properties than S3 (mechanically blended RGO/CNTs-SiCN), which is attributed to the homogeneous dispersion of carbon fillers in S4 and advantage of SSP method. Due to the increased dielectric properties, the composite containing 15 wt.% filler loading also shows outstanding SE value of 67.2 dB, which is the highest value reported for the rGO composites.

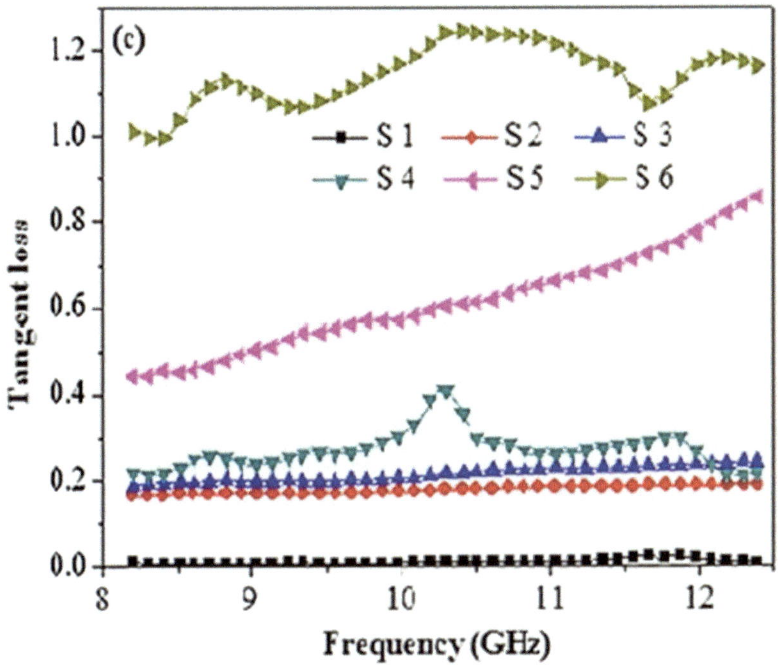

FIGURE 8.9 Tangent loss vs. frequency plot of S1 (neat polysilazane HTT 1800); S2 (SSP-GO-2.0 wt.%); S3 (pyrolyzed product of physically blended GO/CNTs/HTT 1800 mixture); S4 (SSP hybrids with 2 wt.% GO/CNTs); S5 (SSP hybrids with 10.5 wt.% GO/CNTs); S6 (SSP hybrids with 15.0 wt.% GO/CNTs).

Source: Reprinted with permission from Ref. [37]. © 2017 Elsevier.

8.8.4 CEMENT-BASED EMI SHIELDING MATERIALS

Cement is a common structural material in the construction sector with very good environment adaptability features [38]. The need for such construction materials with high EM shielding properties is growing due to the increasing use of electronic devices and gadgets globally. But cement matrix is dielectric; it offers low conductivity [39] and hence shielding-enhancing admixtures that are able to enhance shielding property should be present. Shielding-enhancing admixtures include conductive short fibers like CF, carbon nanotubes (CNT), steel fibers, etc., and conductive particles like coke, carbon black, graphite, graphene, etc. Additionally, there are other conductive filler materials like magnetic woods [40], Ni coated mica [41], Ni plated CF [42], etc. But they are too expensive and possess certain processing difficulties are described in detail below.

8.8.4.1 CARBON-BASED SHIELDING MATERIALS

Recent studies showed that palm oil fuel ash (POFA) which is a rich source of carbon, Fe_2O_3, and other compounds can be used as good microwave absorbers. Ching et al. [43] incorporated POFA into cement which showed a SE of 25.76 dB between 50 MHz and 2 GHz frequency range. Four layers of POFA were created after stirring in water, where the first layer showed better shielding ability. Various characterizations proved that high porosity and carbon content resulted in better shielding. Based on visual observation, layer 1 was floating on the top due to high porosity, layer 2 was the second top layer within submerged layer, layer 3 was the thinnest layer in between layers 2 and 4, and layer 4 was the layer at the bottom. Additionally, energy dispersive x-ray spectroscopy (EDS) were also performed to identify the element content proved that, carbon content of the first layer was the highest among other layers, which accounted to about 85.89%. This content was reduced followed by the sequence of layers. In conclusion, Ching et al. proved that POFA could be used as a potential additive in cement as it blocks about 94.84% of EMF and at the same time reduces the pollution problem of the POFA to the environment.

CNTs were incorporated by Jung et al. [44] in ultra-high performance concrete (UHPC) to study the shielding behavior where they achieved better conductivity and mechanical properties. Dispersion of CNTs was done by sonication and shear mixing along with a superplasticizer. Dispersion was very effective as inferred from the SEM micrographs till the critical incorporation concentration (CIC) and when it increased above the CIC value agglomeration occurred. Figure 8.10(a)–(e) reveals the effectiveness of the dispersion methods for various weight percentages of CNTs. Figure 8.10(a) and (b) show effective dispersion of CNTs in the UHPC matrix while figure c shows numerous CNT fibers between the cracked specimen (fragment sample after the compressive strength test). When the CNT loading increased to 0.8 wt.%, more agglomerations were visible (Figure 8.10(d)–(e)) as they have exceeded the CIC value. In other words, excessive addition of CNTs reduced the relative distance between the fibers, thereby strengthening the Vander Waals forces, ultimately resulting in the agglomeration of CNTs in the matrix. Dispersion was also be correlated with the mechanical properties as per the data revealed. Below the CIC value, dispersion of CNTs enhanced the mechanical properties by pore filling effect, bridging effect and by the formation of a denser C-S-H structure with higher stiffness whereas for values close to CIC, properties weakened. Furthermore, the dispersed CNTs also improved the electrical conductivity of the composite to a percolation threshold value. That is, the SE value also improved with CNT loading due to enhanced electrical conductivity. However, CNT loading above the percolation threshold value did not exhibit the conductivity as the conductive path was already constructed at the percolation threshold.

8.8.4.2 MAGNETIC FILLER-BASED SHIELDING MATERIALS

Zhang et al. [45] prepared two different microwave absorbers that could be used as EM wave shielding materials for buildings. The first one is a single-layer absorber composed of Mn–Zn ferrite, and the second one is a double-layer absorber composed of mortar with silica fume as the surface layer and Mn–Zn ferrite mortar as the loss layer. The reflectivity curves showed that single-layer absorber filled with ferrite had higher reflectivity values than plain cement (PC) samples due to the impedance mismatching characteristics in the single-layer absorber. This was further verified from the fluctuations in the reflection loss curve. The sample with ferrite content 30% showed a smooth curve with maximum reflectivity value below other samples making it as the optimum ferrite content. Furthermore, the double-layer absorber showed excellent absorption property than single-layer absorbers due to a better impedance matching between silica fume (10 wt.%) and ferrite (30 wt.%) in the double-layer absorber.

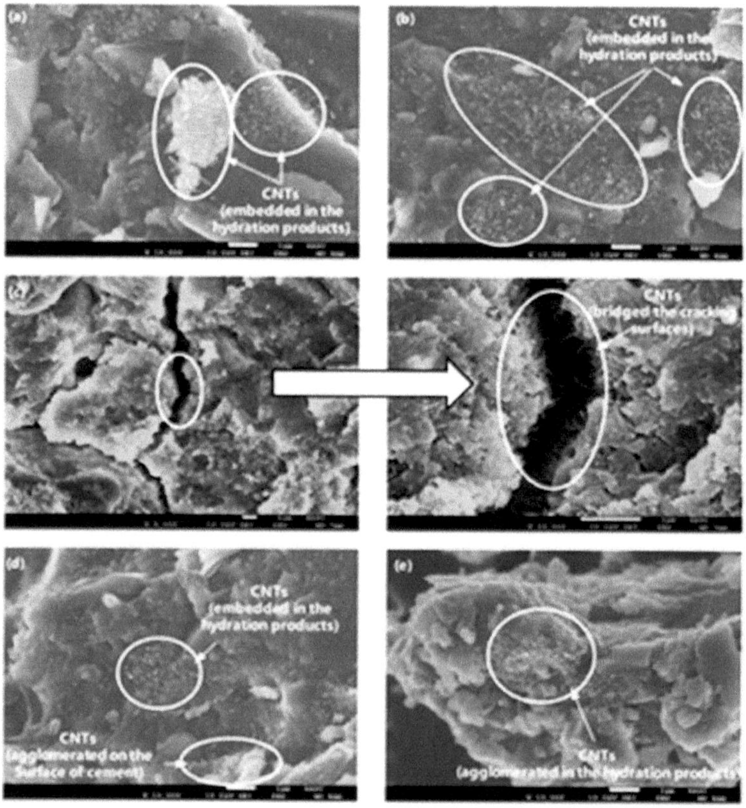

FIGURE 8.10 SEM images of the UHPC-CNT composites: (a) CNT0.2; (b, c) CNT0.5; (d) CNT0.8; and (e) CNT1.0.

Source: Reprinted with permission from Ref. [44]. © 2020 Elsevier.

Yao et al. [46] synthesized a novel EMI shielding material from nickel fiber-based cement composites utilizing silica fume and colloidal graphite. The effect of silica fume and colloidal graphite content on shielding properties was studied and a SE value greater than 40 dB were obtained. Initially, the effect of Ni doping on electrical conductivity and SE were analyzed which revealed that Ni doping improved those values by several orders of magnitude when compared with the sample without Ni doping. Synergistic effect of silica fume (0 to 30 vol%) and 1 vol% Ni fiber on electrical conductivity and SE value were also analyzed. Result reveal that the shielding property is best at 20 vol% silica fume doping with an average SE value of 14.53 dB whereas the composite devoid of silica fume doping showed only 11.53 dB. Similarly, electrical conductivity values were best at 20 vol% silica fume doping with conductivity value of 7.09 × 10^{-4} s.cm^{-1}. Studies based on the synergistic effect of colloidal graphite (0 to 30 vol%) and 1 vol% Ni fiber showed that the SE and electrical conductivity values are the best at 20 vol% colloidal graphite content with SE and electrical conductivity values of 35.59 dB and 7.25 × 10^{-4} s.cm^{-1}, respectively. Furthermore, the synergistic effect of varying Ni fiber (1, 3, 5, 7, 9 vol%) with silica fume (20 vol%) and colloidal graphite on SE were also investigated. Composite with Ni fiber (9 vol%) and silica fume showed a SE and conductivity value of 35.50 dB and 2.86 × 10^{-3} s.cm^{-1}, respectively. Whereas, composite with 5 vol% colloidal graphite and silica fume showed a SE and conductivity of 51.19 dB and 2.95 × 10^{-3} s.cm^{-1}, respectively showing the obvious effect of colloidal graphite on SE. Further increasing the Ni content resulted in a decrement in the SE value due to Ni fiber aggregation.

8.8.4.3 DIELECTRIC FILLER-BASED SHIELDING MATERIALS

Cement-based ceramic pellets with MnO_2 as filler materials were prepared by wet mixing method for shielding applications [47]. The shielding materials were made of Portland cement with the addition of 0.1, 0.5, 1, 5, and 10 wt.% of MnO_2. The pellets were sintered at 850°C for 5 h and then polished prior to characterizations. Investigations reveal that the MnO_2–cement pellets have good dielectric properties, i.e., a high dielectric constant value of ~300 and low dielectric loss value <0.3, which determines the electrical characteristics of a material. The dielectric constant value increased with increasing filler content in the cement matrix, which is attributed to the magnetic properties of MnO_2, improvement in density and porosity. The SE values of the MnO_2–cements fluctuated between 2 dB and 9 dB in the frequency range of 8–13 GHz as shown in Figure 8.11. The sample with 10 wt.% MnO_2 had SE values of up to 9 dB. Most of the samples with high additive concentrations had SE greater than 7 dB above 12 GHz frequency range, thus inferring that presence of MnO_2 plays a crucial role in improving the SE of cement composite.

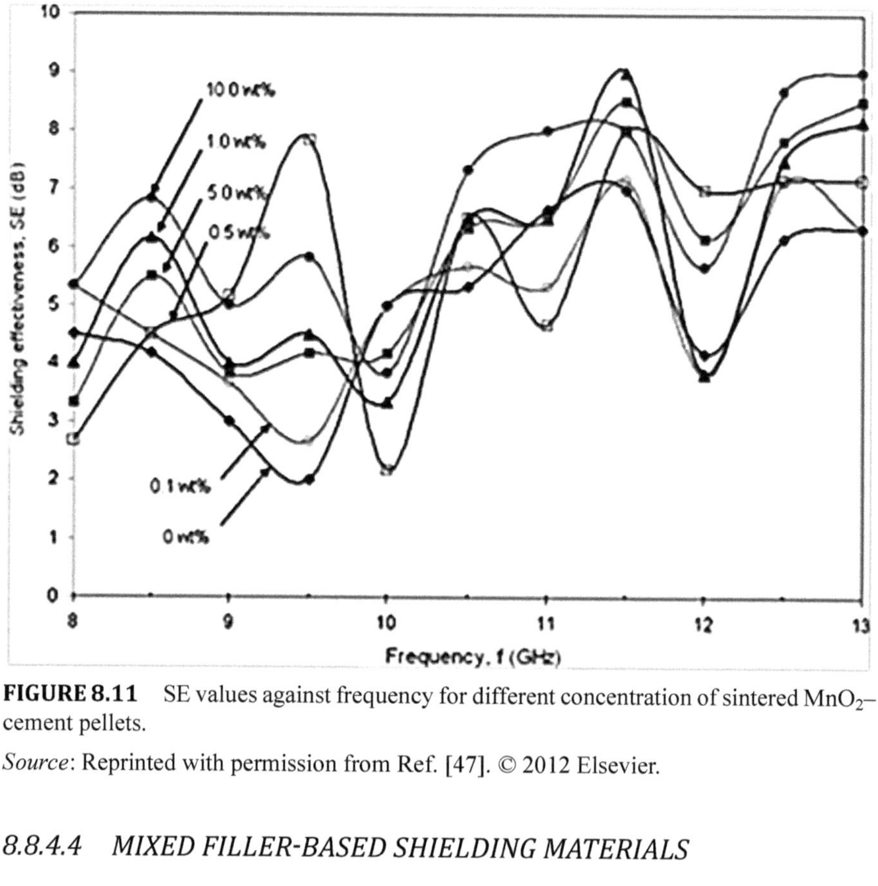

FIGURE 8.11 SE values against frequency for different concentration of sintered MnO_2–cement pellets.

Source: Reprinted with permission from Ref. [47]. © 2012 Elsevier.

8.8.4.4 MIXED FILLER-BASED SHIELDING MATERIALS

Liu et al. [48] developed cement-based composite by homogeneously dispersing carbon fiber and Fe_2O_3 nanoparticles into the cement matrix. The SE of composites enhanced with the increase in Fe_3O_4 loading. For 0.4 wt.% CF and 5 wt.% Fe_3O_4 loading, the SE value reached 29.8 dB in the X band frequency range, which is attributed to the synergistic effect between CF and Fe_3O_4 nanoparticles due to the strong conduction loss and multiple scattering. Additionally, this composite showed 34.4% enhancement in SE value than the CF/cement composite making them potential materials for EM shielding.

Medicinal package wastes like blisters (a composite of Aluminum and LDPE) were recently used as filler materials in cement for low profile EMI shielding applications. Suravarjhula et al. [49] prepared cement/blister composite using a mortar mixer and analyzed the SE values in J and X band. Ordinary cement mortar using sand and water for plastering purposes and cement mortar containing blister particles were subjected to SE experiment. Studies revealed that the LDPE part in the blister provided mechanical strength and the Al part imparted an additional SE value of 6 dB. The composite were also subjected to mechanical property studies and no appreciable degradation of compressive strength was observed. Thus, this composite proposes a solution for blister disposal and at the same time provided a low cost, field-deployable EM shield at the structural level.

8.8.5 TEXTILE-BASED EMI SHIELDING MATERIALS

Conventional materials used for shielding applications are metallic materials and metal-coated plastics with high EM properties. But, possess certain disadvantages related to weight, flexibility, and cost-effectiveness, which has re-routed the attention towards textiles for EMI shielding application. When compared with conventional shielding materials, the textile-based shields have the advantages of durability, flexibility, adaptability, low cost, and lightweight nature. Different type of fabrics used includes metalized fabric, nonwoven fabric, knitted fabric, conductive polymer coated fabric, etc. Some of the commercial products include Phantom fabric, NaturaShield, FlecTron-N, etc. [50].

8.8.5.1 METALLIZED FABRICS AS SHIELDING MATERIALS

Sim et al. [51] developed self-healable and flexible GO/silver nanowire films and textiles with excellent EMI shielding performance. The maximum SE value of 92 dB was recorded for an 18 μm film thickness (refer to Figure 8.12(a)). The composite film and coated textile were tolerant of damage induced by cracking and scratching. The cracking process reduced the electrical conductivity, mechanical properties and the SE value was reduced from 72 dB to 56 dB at 8.2 GHz frequency range. After the healing process, the SE value reached 71 dB and mechanical properties were restored. The SE value of textile reached a maximum of 30 dB which is suitable for commercial EMI shielding application. In addition, the textile exhibited good flexibility, and showed excellent mechanical stability with no change in shielding performance (Figure 8.12(b)) after 1,000 bending cycles.

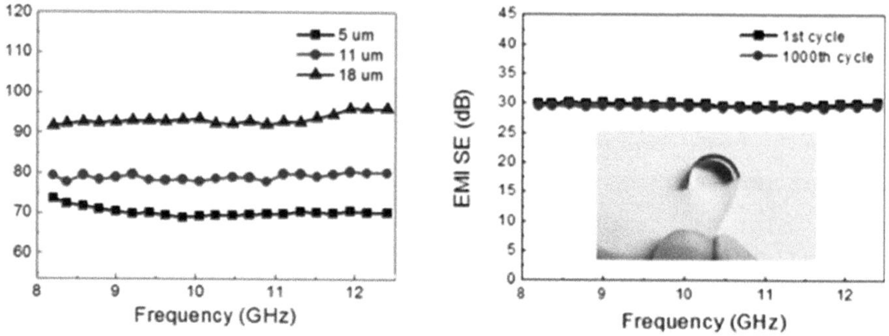

(a) (b)

FIGURE 8.12 (a) EMI SE value of the composite film with Ag nanowire content of 80 wt.% and at different film thicknesses; (b) EMI SE of EMI shielding textile with repeated bending cycles [inset image show the bent EMI shielding textile].

Source: Reprinted with permission from Ref. [51]. © 2019 Elsevier.

Han et al. [52] developed shielding textile as protective clothes from PET fabric using electroless Cu-plating technique. The plating technique consisted of many steps like scouring, rinsing, etching, catalyzation, acceleration, Cu-plating, rinsing, post-treatment for stopping Cu-reduction, and drying. All the pre-treatments were studied in relation to the SE values and physical properties of treated fabrics. A high SE value of approx. 55 dB were obtained when scoured with 0.3% NaOH, and 1% Triton X-100, and etched with HCl and HNO_3.

8.8.5.2 CARBONACEOUS FILLERS-BASED SHIELDING MATERIALS

GO dyed cotton fabrics were developed by Islam et al. [53] through a continuous dyeing method which was later reduced by ascorbic and then modified with dopamine. The FESEM (Figure 8.13(a)–(c)) results showed significant improvements for the adhesion of rGO uniformly on the fibre surface. After 20 dyeing cycles the rGO content increased, which in turn increased the conductivity values from 1.2×10^{-7} S/cm^{-1} to 2.3×10^{-1}S/cm^{-1} and SE values from −26 to −35 dB in the range of 30–1,530 MHz in X-band. Furthermore, the rGO coated textile also showed improvement in

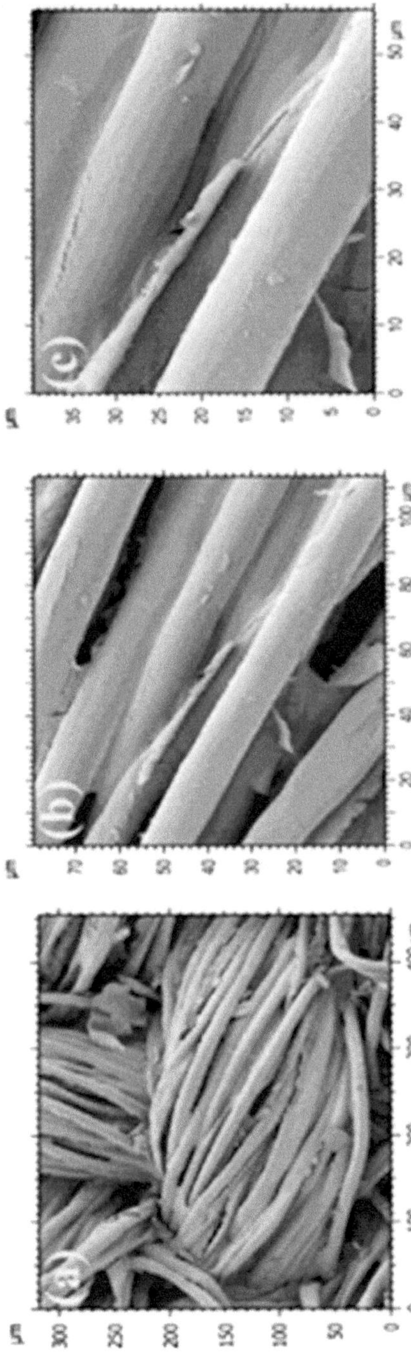

FIGURE 8.13 (a) FESEM micrographs of rGO dyed cotton fabric at 400 μm; (b) 100 μm; (c) 50 μm.

Source: Reprinted with permission from Ref. [53]. © 2019 Elsevier.

tensile strength. The tensile strength increased to 27–30% after five dyeing cycles, which further improved up to 60–68% after 15–20 dyeing cycles.

ZnO and rGO coated wearable cotton fabrics have been prepared by Gupta et al. [54] for EMI shielding in the X-band frequency range. The coating of ZnO nanoparticles was achieved by an in-situ sol-gel method, whereas the rGO coating was achieved by spraying GO solution followed by thermal reduction as shown in Figure 8.14. The results inferred that the SE_T increased with the rGO loading, while, the reflection loss decreased. The rGO/ZnO coated cotton loaded with 7 wt.% rGO achieved highest SE_T of 54.7 dB (~99.999%), which is attributed to 17.783% of reflection and 82.216% of absorption which is due to the dielectric nature of ZnO nanoparticles, conductive nature of rGO sheets and the core-shell structure of coated cotton fabric. It was also found that very high loading of rGO blocked all the pores of the cotton fabric hindering the absorption of the EM waves. Additionally, it was also found that the rGO/ZnO coated cotton fabrics can also be used in different applications such as antibacterial fabrics, oil/ water separation, counter electrode in dye sensitize solar cells, and electrochemical capacitors.

8.8.5.3 INTRINSICALLY CONDUCTING POLYMER (ICP)-BASED SHIELDING MATERIALS

Gahlout et al. [55] impregnated PPy by in-situ polymerization onto various fabrics like cotton, polyester, nylon, and cotton lycra and analyzed their EMI shielding properties. Four impregnation steps were carried out to decrease the springiness of the fabrics. Conductivity measurements revealed that as the number of impregnation cycles increased (refer Figure 8.15), a regular conducting network is created thereby increasing the conductivity. Lycra fabric exhibited the highest shielding value of 18.7 dB at 0.3 mm thickness which is in good agreement with the conductivity values of this fabric. This SE value increased further to 25 dB upon adhering two layers in the X band frequency range. PPy impregnated PET fabric showed the second-highest SE value of 10 dB while cotton showed the least SE value around 4 dB. Furthermore, analysis on thermal stability showed that lycra fabric had the lowest initial degradation temperature among all impregnated fabrics due to highest add-on of PPy and the tan δ value obtained was high (15.6–8.7) among the fabrics because of high loading of PPy over lycra fabric.

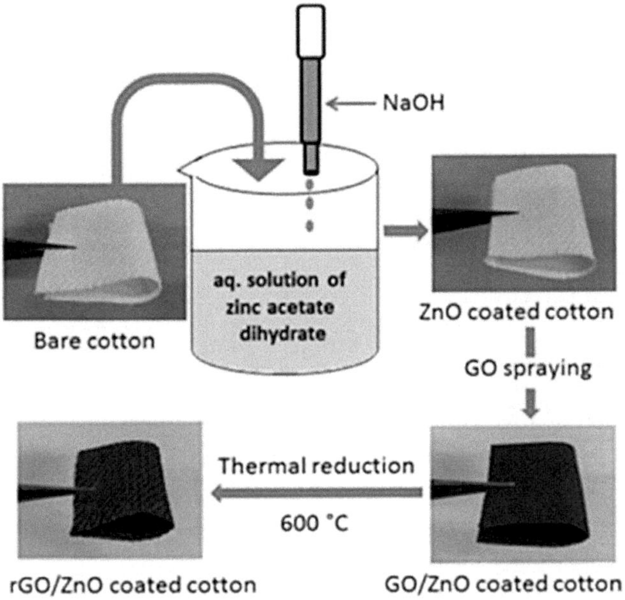

FIGURE 8.14 Schematic representation for the preparation of rGO/ZnO coated cotton fabric.

Source: Reprinted with permission from Ref. [54]. © 2020 Elsevier.

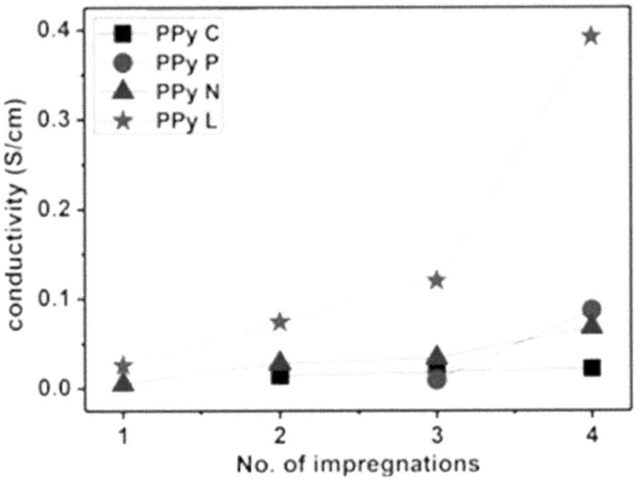

FIGURE 8.15 Variation in conductivity with increasing impregnation cycles. PPy L, N, P, and C represent lycra, nylon, polyester, and cotton fabric impregnated with PPy, respectively.

Source: Reprinted with permission from Ref. [55]. © 2019 Elsevier.

PANI-coated nylon lycra fabric with strain sensing and EMI shielding features have been developed by Muthukumar et al. [56] by an in situ chemical oxidation polymerization process. The developed PANI-coated fabric exhibited a SE value of 26 dB in the 8 GHz to 12 GHz frequency range. Additionally, the measurement of conductivity change with strain showed that the fabrics exhibited high strain sensitivity. In addition to high sensitivity, environmental stability was also better.

8.9 CONCLUSION

This review comprehensively highlights various materials for EMI shielding, particularly textiles, cement, ceramics, and polymers. Sources, effects, factors affecting EMI shielding, and different fillers used were discussed extensively. All the studies revealed that polymers, their hybrids, multi-structural composites, and blends are very promising for EMI shielding applications. Among the various materials discussed, polymers are found to be potential materials for shielding applications as they impart a lightweight and high EM wave absorption properties. Despite considerable advancements in this field, numerous limitations in this field remain. For instance, as conventional polymers are insulating in nature, and if no sufficient conducting network is formed by conductive fillers, it leads to skin depth enlargement for attenuating EM waves. Many researchers have reported lightweight foams as efficient EM wave absorbers, but factors that affect bandwidth, skin depth, and the effect of temperature on EM properties are not studied in detail. We conclude that lightweight multicomponent structures and 3D structures of polymeric materials impart very good mechanical properties, thermal stability, and EM wave attenuation properties.

KEYWORDS

- **carbon nanotubes**
- **composites**
- **electromagnetic interference**
- **ferrites**
- **graphene**
- **radio frequency**
- **shielding efficiency**

REFERENCES

1. Genius, S. J., (2008). *Public Health, 122*, 113.
2. Erogul, O., Oztas, E., Yildirim, I., Kir, T., Aydur, E., Komesli, G., Irkalita, H. C., et al., (2006). *Arch. Med. Res., 37*, 840.
3. Chung, D. D. L., (2020). Materials for electromagnetic interference shielding. *Materials Chemistry and Physics*, 123587.
4. Joshi, A., & Suwarna, D., (2015). Carbon nanostructure composite for electromagnetic interference shielding. *Pramana, 84*(6), 1099–1116.
5. Chung, D. D. L., (2001). Electromagnetic interference shielding effectiveness of carbon materials. *Carbon, 39*(2), 279–285.
6. Jaroszewski, M., Sabu, T., & Ajay, V. R., (2018). *Advanced Materials for Electromagnetic Shielding: Fundamentals, Properties, and Applications*. John Wiley & Sons.
7. Ozdemir, F., & Aysegul, K., (2011). Electromagnetic waves and human health. *Electromagnetic Waves*. IntechOpen.
8. Aghaei, M., et al., (2012). A review on the impact of the electromagnetic radiation (EMR) on the human's health. *Proceedings National Graduate Conference (NatGrad2012)*. Universiti Tenaga Nasional, Putrajaya Campus.
9. Barsam, T., et al., (2012). Effect of extremely low frequency electromagnetic field exposure on sleep quality in high voltage substations. *Iranian Journal of Environmental Health Science & Engineering, 9*(1), 15.
10. Kolmodin-Hedman, B., et al., (1988). Health problems among operators of plastic welding machines and exposure to radiofrequency electromagnetic fields. *International Archives of Occupational and Environmental Health, 60*(4), 243–247.
11. Bini, M., et al., (1986). Exposure of workers to intense RF electric fields that leak from plastic sealers. *Journal of Microwave Power and Electromagnetic Energy, 21*(1), 33–40.
12. Shukla, V., (2019). Review of electromagnetic interference shielding materials fabricated by iron ingredients. *Nanoscale Advances, 1*(5), 1640–1671.
13. Wang, J., et al., (2016). Combined use of lightweight magnetic Fe_3O_4-coated hollow glass spheres and electrically conductive reduced graphene oxide in an epoxy matrix for microwave absorption. *Journal of Magnetism and Magnetic Materials, 401*, 209–216.
14. Yousefi, N., et al., (2014). Highly aligned graphene/polymer nanocomposites with excellent dielectric properties for high-performance electromagnetic interference shielding. *Advanced Materials, 26*(31), 5480–5487.
15. Phan, C. H., Mariatti, M., & Koh, Y. H., (2016). Electromagnetic interference shielding performance of epoxy composites filled with multiwalled carbon nanotubes/manganese zinc ferrite hybrid fillers. *Journal of Magnetism and Magnetic Materials, 401*, 472–478.
16. Arora, M., Wahab, M. A., & Parveen, S., (2014). Permittivity and electromagnetic interference shielding investigations of activated charcoal loaded acrylic coating compositions. *Journal of Polymers, 2014*.
17. Zhou, H., et al., (2020). Conductive and fire-retardant wood/polyethylene composites based on a continuous honeycomb-like nanoscale carbon black network. *Construction and Building Materials, 233*, 117369.

18. Yuan, D., et al., (2020). Recyclable conductive epoxy composites with segregated filler network structure for EMI shielding and strain sensing. *Composites Part A: Applied Science and Manufacturing, 132*, 105837.
19. Joshi, A., et al., (2013). Graphene nanoribbon–PVA composite as EMI shielding material in the X band. *Nanotechnology, 24*(45), 455705.
20. La Rosa, A. D., et al., (2020). Recovery of electronic wastes as fillers for electromagnetic shielding in building components: An LCA study. *Journal of Cleaner Production,* 124593.
21. Chen, Y., et al., (2020). Multifunctional cellulose/rGO/Fe_3O_4 composite aerogels for electromagnetic interference shielding. *ACS Applied Materials & Interfaces, 12*(19), 22088–22098.
22. Singh, R., & Suresh, G. K., (2014). Nanocomposites based on transition metal oxides in polyvinyl alcohol for EMI shielding application. *Polymer Bulletin, 71*(2), 497–513.
23. Vaid, K., et al., (2020). Study of dielectric and electromagnetic shielding behavior of $BaTiO_3$-$CoFe_2O_4$ filled LDPE composite. *Polymer Composites*.
24. Wu, Y., et al., (2017). Ultralight graphene foam/conductive polymer composites for exceptional electromagnetic interference shielding. *ACS Applied Materials & Interfaces 9*(10), 9059–9069.
25. Gupta, N. D., et al., (2011). Kinetics of micelle formation and their effect on the optical and structural properties of polypyrrole nanoparticles. *Colloids and Surfaces A: Physicochemical and Engineering Aspects, 385*(1–3), 55–62.
26. Omastova, M., et al., (2003). Synthesis and structural study of polypyrrole prepared in the presence of surfactants. *Synthetic Metals, 138*(3), 447–455.
27. Parit, M., et al., (2020). Polypyrrole and cellulose nanofiber based composite films with improved physical and electrical properties for electromagnetic shielding applications. *Carbohydrate Polymers,* 116304.
28. Agrawal, R., et al., (2020). Significantly high electromagnetic shielding effectiveness in polypyrrole synthesized by eco-friendly and cost-effective technique. *Journal of Applied Polymer Science, 137*(48), 49566.
29. Pande, A., et al., (2017). Electromagnetic shielding behavior of polyaniline using red mud (industrial waste) as filler in the X–band (8.2–12.4 GHz) frequency range. *Materials Chemistry and Physics, 189*, 22–27.
30. Saini, P., et al., (2013). High permittivity polyaniline–barium titanate nanocomposites with excellent electromagnetic interference shielding response. *Nanoscale, 5*(10), 4330–4336.
31. Li, Y., Huang, Y., Qi, S., Niu, L., Zhang, Y., & Wu, Y., (2012). Preparation, magnetic and electromagnetic properties of polyaniline/strontium ferrite/multiwalled carbon nanotubes composite. *Appl. Surf. Sci., 258*, 3659–3666.
32. Zeng, S., et al., (2020). From waste to wealth: A lightweight and flexible leather solid waste/polyvinyl alcohol/silver paper for highly efficient electromagnetic interference shielding. *ACS Applied Materials & Interfaces*.
33. Hongxing, et al., (2017). Microstructures and EMI shielding properties of composite ceramics reinforced with carbon nanowires and nanowires-nanotubes hybrid. *Ceramics International, 43*(15), 12221–12231.

34. Barani, Z., et al., (2020). Multifunctional graphene composites for electromagnetic shielding and thermal management at elevated temperatures. *Advanced Electronic Materials,* 2000520.
35. Raghvendra, S. Y., et al., (2019). $NiFe_2O_4$ nanoparticles synthesized by dextrin from corn-mediated sol-gel combustion method and its polypropylene nanocomposites engineered with reduced graphene oxide for the reduction of electromagnetic pollution. *ACS Omega 4*(26), 22069–22081.
36. Qing, Y., et al., (2018). $NiFe_2O_4$ nanoparticles filled $BaTiO_3$ ceramics for high-performance electromagnetic interference shielding applications. *Ceramics International, 44*(7), 8706–8709.
37. Liu, X., et al., (2017). Single-source-precursor derived RGO/CNTs-SiCN ceramic nanocomposite with ultra-high electromagnetic shielding effectiveness. *Acta Materialia, 130*, 83–93.
38. Guan, H., et al., (2006). Cement based electromagnetic shielding and absorbing building materials. *Cement and Concrete Composites, 28*(5), 468–474.
39. Zornoza, E., et al., (2020). Addition of carbon nanofibers to cement pastes for electromagnetic interference shielding in construction applications. *Nanotechnology in Cement-Based Construction* (pp. 377–391). Jenny Stanford Publishing.
40. Oka, H., & Hiromitu, F., (1999). Experimental study on magnetic and heating characteristics of magnetic wood. *Journal of Applied Physics, 85*(8), 5732–5734.
41. Jiang, G., et al., (2002). Preparation of nickel coated mica as a conductive filler. *Composites Part A: Applied Science and Manufacturing, 33*(5), 745–751.
42. Lu, G., Xiaotian, L., & Hancheng, J., (1996). Electrical and shielding properties of ABS resin filled with nickel-coated carbon fibers. *Composites Science and Technology, 56*(2), 193–200.
43. Ching, N. O. L., et al., (2017). Exploring the potential of palm oil fuel ash (POFA) in EMI shielding effectiveness. *J. Mech. Eng., 2*, 101–111.
44. Jung, M., et al., (2020). Carbon nanotubes (CNTs) in ultra-high performance concrete (UHPC): Dispersion, mechanical properties, and electromagnetic interference (EMI) shielding effectiveness (SE). *Cement and Concrete Research, 131*, 106017.
45. Zhang, X., & Wei, S., (2010). Microwave absorbing properties of double-layer cementitious composites containing Mn–Zn ferrite. *Cement and Concrete Composites, 32*(9), 726–730.
46. Yao, W., et al., (2017). Effect of silica fume and colloidal graphite additions on the EMI shielding effectiveness of nickel fiber cement based composites. *Construction and Building Materials, 150*, 825–832.
47. Hutagalung, S. D., et al., (2012). Effect of MnO_2 additive on the dielectric and electromagnetic interference shielding properties of sintered cement-based ceramics. *Ceramics International, 38*(1), 671–678.
48. Liu, Z., et al., (2017). Enhanced electromagnetic interference shielding of carbon fiber/cement composites by adding Ferro ferric oxide nanoparticles. *Construction and Building Materials, 151*, 575–581.
49. Suravarjhula, V. K., et al., (2018). Cement based composite loaded with medicinal package waste for low profile electromagnetic shielding. *2018 USNC-URSI Radio Science Meeting (Joint with AP-S Symposium)*. IEEE.

50. Maity, S., et al., (2013). Textiles in electromagnetic radiation protection. *Journal of Safety Engineering, 2*(2), 11–19.
51. Sim, H. J., et al., (2019). Self-healing graphene oxide-based composite for electromagnetic interference shielding. *Carbon, 155*, 499–505.
52. Han, E. G., Eun Ae, K., & Kyung, W. O., (2001). Electromagnetic interference shielding effectiveness of electroless Cu-plated PET fabrics. *Synthetic Metals, 123*(3), 469–476.
53. Islam Md, Z., et al., (2019). Continuous dyeing of graphene on cotton fabric: Binder-free approach for electromagnetic shielding. *Applied Surface Science, 496*, 143636.
54. Gupta, S., et al., (2020). Reduced graphene oxide/zinc oxide coated wearable electrically conductive cotton textile for high microwave absorption. *Composites Science and Technology, 188*, 107994.
55. Gahlout, P., & Veena, C., (2019). Microwave shielding behavior of polypyrrole impregnated fabrics. *Composites Part B: Engineering, 175*, 107093.
56. Muthukumar, N., Thilagavathi, G., &. Kannaian, T., (2015). Polyaniline-coated Nylon Lycra fabrics for strain sensor and electromagnetic interference shielding applications. *High Performance Polymers, 27*(1), 105–111.

CHAPTER 9

Refabrication of Natural Polymeric Material Through Grafting: A Review

SWETA SINHA

Assistant Professor, Department of Applied Sciences, Amity University Jharkhand, Ranchi, Jharkhand, India, E-mail: sweta.sinha2203@gmail.com

ABSTRACT

Growing environmental degradation has led to an increased demand for hybrid materials based on renewable biological resources for their tailor-made applications. However, the properties of natural polymers are limited, so their chemical modification is essential for the improvement of the desired properties. Chemical modification of natural polymeric materials through grafting is a key method to synthesize new hybrid materials for selective desirable properties. Grafting enhances the compatibility between natural and synthetic polymers. Graft copolymerization of natural polymer is a simple alternative to a number of problems since it expands the spectrum of applications and introduces new possibilities. Commonly, Grafting takes place through free radical polymerization on preformed polymer or a monomeric unit of another polymer. Free radicals are generated on the backbone of biopolymers either by chemical-induced methods (using a chemical-free radical initiator) or radiation-induced methods. This research emphasizes the use of several chemical initiators and irradiation techniques in the synthesis of graft copolymers.

Advances in Diverse Applications of Polymer Composites: Synthesis, Application, and Characterization.
Suji Mary Zachariah, Yang Weimin, Maciej Jaroszewski, & Sabu Thomas (Eds.)

9.1 INTRODUCTION

Polymeric Materials play significant roles in controlling and minimizing pollution and contamination in the environment. The role of polymer based materials in decreasing the toxicity of exhausts, industrial waste, contaminants, and different recycling materials including water is currently the area of intensive research activities.

Rapid population rise and massive industrial growth in the world have engendered various environmental problems and health hazards for the mankind. Hence, deterioration in environmental conditions and depletion of natural resources necessitate scientists to develop and improve the quality of materials from renewable resources so as to reduce various environmental pollutants.

Increased environmental awareness, growing demands for green materials (materials that have a significant environmental impact in our lives), and a greater interest in natural materials for the production of conventional and advanced polymer materials have culminated from technological innovation, increased levels of education, and innovative work in the field of environment. Chemical modification is required to improve desired attributes such as water absorbency, ion-exchange capacities, elasticity, mechanical strength, and heat resistance because the existing condition of these natural/nature-derived materials limits their features. Modifications of polymer can convert a soluble polymer into an insoluble one or vice versa.

9.2 MATERIALS AND METHODS

9.2.1 GRAFT COPOLYMER

Robust research in the synthesis of materials derived from renewable resources [1, 2] has led to the genesis of novel hybrid materials. Synthesis of these hybrid materials is an essential step to reduce environmental impact and develop new applications for renewable resource-derived materials. Growing environmental degradation has led to an increased demand of the hybrid materials based on renewable biological resources (e.g., Casein) for various industrial, biomedical, and food applications. Although a lot of investigations are evident in polysaccharide and protein front [5–7].

Modification of biopolymer has been investigated in many different circumstances, both for the production of new renewable resource derived materials and for industrial and biomedical applications [8–11].

9.2.1.1 GRAFT COPOLYMERIZATION AND ITS ADVANTAGES

Graft copolymer is a special type of branched copolymers containing a long chain of one monomer called preformed or backbone polymer and branch/graft of long sequence of usually another monomer [12, 13]. Grafted strings (branches) usually have a lower degree of polymerization than the main chain (preformed polymer). Usually, the Grafted chains are randomly distributed all along the backbone of preformed polymer, although recent advances in synthetic techniques allow the preparation of enhanced defined structures [13]. Figure 9.1 [14] illustrates the simplest form of a graft copolymer.

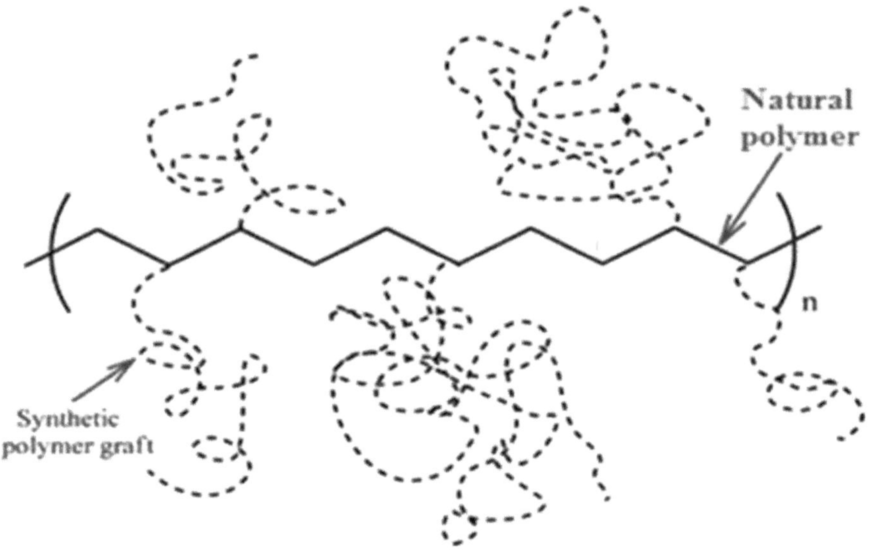

FIGURE 9.1 Concept of graft copolymer.

Further, in the presence of crosslinking agent (e.g., MBA) graft co-polymers form 3-dimensional cross-linked structure having the ability to absorb a very large amount of water [15]. Various factors such as monomer concentration, ionic strength, pH-value, amount of crosslinker mainly influence the extent of crosslinking made via the free radical polymerization mechanism [15, 16].

Graft copolymerization techniques include free radical polymerization, ionic polymerization, redox polymerization, coordination polymerization and atomic transfer radical polymerization. Other approaches are less commonly used than free radical polymerization. This process is still used to make about 60% of all accessible polymers [17].

Aqueous and heterogeneous media are commonly used for grafting. In an aqueous medium, the amorphous and accessible regions of preformed polymer swell, whereas monomers diffuse, initiating the grafting reactions [18]. Due to the following attractive characteristics of grafting, the free radical polymerization takes place [19]:

- When compared to competing technologies, it is inexpensive and easy to apply;
- It is suitable for the ample range of monomers to the polymerization;
- It takes place in aqueous media with a good yield of the final product.

A set of highly reactive, short-lived reaction components with one or more unpaired electrons is referred to as a "free radical." Additional polymers are formed by free radical polymerization involving the following three steps:

- Generation of reactive sites on the backbone of preformed polymer;
- Addition of monomer to the polymeric backbone; and
- Propagation of monomeric units.

Free radicals generated on the backbone of biopolymers either by chemical-induced methods (using chemical-free radical initiator) or radiation-induced methods [20]. Crosslinking between same/different polymers uses the same chemical mechanism as grafting [21]. The reaction mechanism of free radical graft copolymerization is depicted in Figure 9.2 [21, 22].

The above polymerization techniques can be obtained by three general approaches namely: (i) *grafting-onto*, in which side chains are preformed and then attached to the backbone; (ii) *grafting-through*, in which the macromonomers are copolymerized; and (iii) *grafting-from*, in which the monomer is grafted from the backbone:

1. **Graft Copolymers by *Grafting onto* Method:** In *grafting onto* method, a reactive functional group on the polymer backbone is

coupled with a preformed polymer chain containing a suitable reactive functional group on its end (Figure 9.3(a)). Ionic polymerization methods are the most suitable for the preparation of the graft unit, since polymers with controlled structures and reactive chain-end functionality can be prepared [22]. This method is fairly controllable.

2. **Graft Copolymer by *Grafting through* Method:** Grafting occurs when a monomer polymerizes in the presence of a polymer with pendant unsaturated groups (Figure 9.3(b)). However, Cross-linking occurs when unsaturation from two separate backbones is incorporated into a developing site [23].
3. **Graft Copolymers by *Grafting from* Method:** The procedure is carried out once the backbone has been prepared. On the main chain, free radicals are produced, which can polymerize the monomers that will make up the branches [23]. The *grafting from* is one of the most often utilized techniques (Figure 9.3(c)).

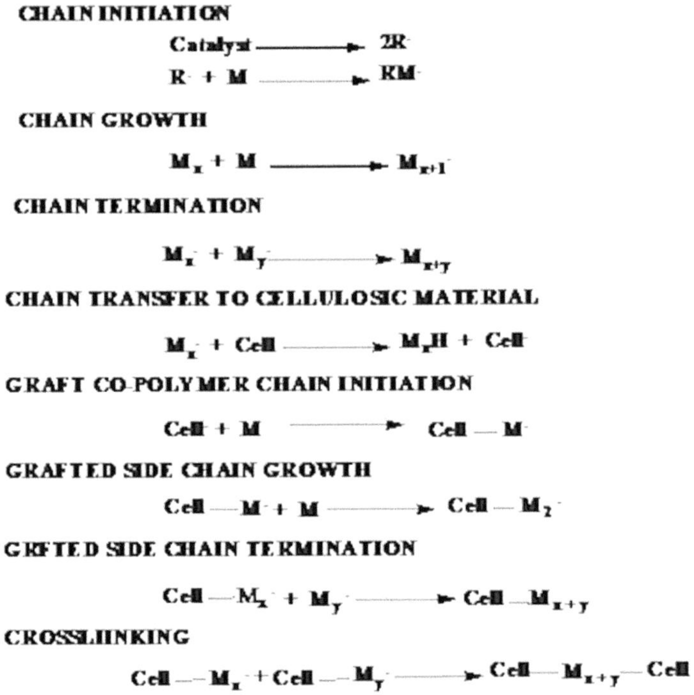

FIGURE 9.2 Reaction mechanism for the formation of grafted and crosslinked material.

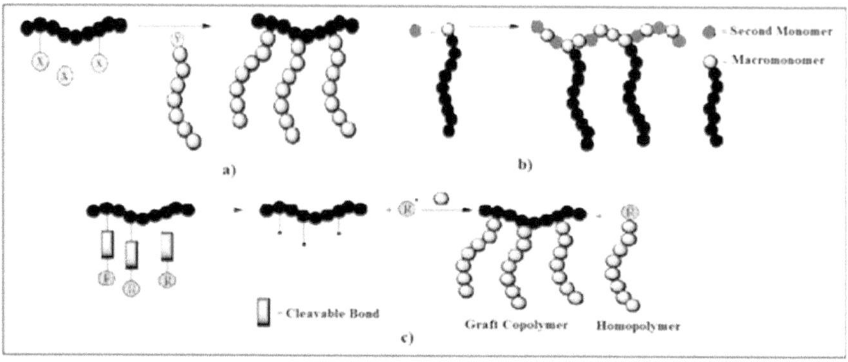

FIGURE 9.3 Synthesis of graft copolymer by (a) *grafting onto*; (b) *grafting through*; and (c) *grafting from* methods.

Consequently, the free radicals generated on the main chain for the synthesis of graft copolymer and crosslinked materials can be achieved by various techniques such as Conventional free radical graft copolymerization (using chemical-free radical initiator) [24] and Radiation-induced graft copolymerization [25, 26] (Figure 9.4).

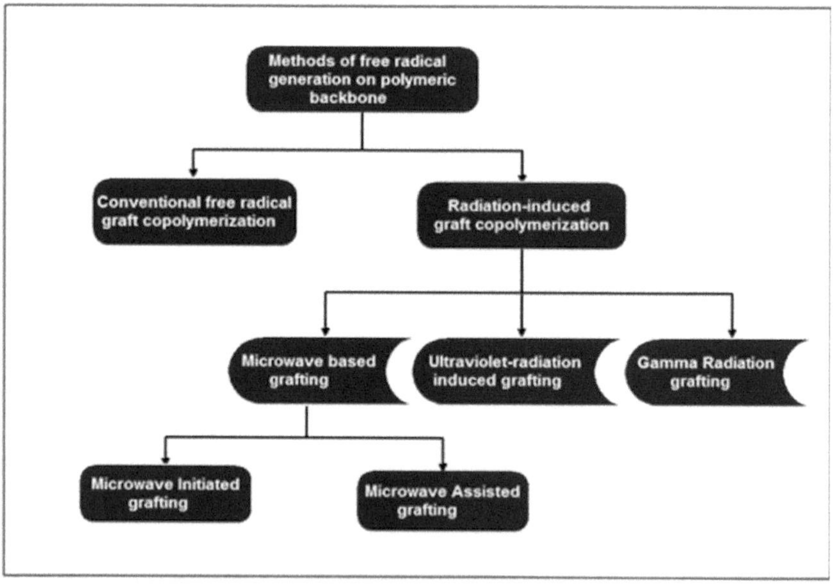

FIGURE 9.4 Diagram showing various methods of creating reactive sites on polymeric backbone.

9.2.2 CONVENTIONAL FREE RADICAL GRAFT COPOLYMERIZATION

Grafting multiple monomers onto an existing polymer (Preformed polymer/Backbone polymer) using a chemical-free radical initiator is a common approach. During the process of polymerization, the radical formation can occur either on the backbone polymer or on the monomer to be grafted. If the radical formation is on monomer, then Homopolymer is produced. Therefore, radical formation at various sites on the backbone is preferred. The various chemical-free radical initiators utilized for creating reactive sites on the preformed polymer are mentioned below [27]:

- Redox reagents (e.g., Fenton's reagent (Fe^{+2}/H_2O_2), persulfate, peroxides);
- Metal chelates (e.g., $VO(acac)_2$, $Fe(acac)_3$);
- Transition metal ions (e.g., Cr^{6+}, Ce^{4+}, Co^{3+}, V^{5+});
- Direct oxidation using Ce (IV) ions as free radical initiator.

For graft copolymerization onto biopolymers, the Ceric ion initiating approach has been widely applied [28, 29]. The Ce (IV) ion (oxidant) formed a complex compound with the biopolymer's functional groups (-OH, $-NH_2$). The resulting complex produces disproportionately reactive regions on the backbone of the biopolymer. Several studies [30, 31] have provided evidence of a complex formation between Ce (IV) and cellulose. The principles of the reaction are illustrated in Figure 9.4.

The Ceric ion initiation method has been used most widely for initiating grafting of monomers (acrylamide (AAm), acrylonitrile (AN), methyl methacrylate (MMA)) onto natural polymeric backbone (cellulose) [30, 31].

Free radical initiator induced grafting processes, on the other hand, are quite difficult and cannot be scaled up beyond bench scale. Furthermore, the graft copolymer produced by this method has a low %grafting reproducibility.

9.2.3 RADIATION-INDUCED GRAFT COPOLYMERIZATION

In this method, reactive sites generate on the biopolymer using high-energy radiations (γ-rays, UV-, and microwave radiations). Number of monomers has been grafted to most of the commercially accessible addition and condensation polymers. Although high-energy radiation-induced grafting is very much suitable for commercial/bulk production of the graft copolymer, but these high-energy radiations also cause unpredictable damage (radiolysis) of the backbone polymer, making the graft copolymer prone to failure [32, 33]. Due to their exceptional stability, the processes are especially suitable for fluoropolymers [34–37].

In the presence of a photoinitiator such as benzophenone or benzoin, UV radiation-induced grafting can also be utilized to activate grafting [38]. It's best for surface grafting because of its minimal penetration capabilities, which results in uneven and low-percentage grafting.

9.2.3.1 MICROWAVE-BASED TECHNIQUE: A QUICK APPROACH OF GRAFTING

A recent method for creating reactive sites on the polymeric backbone uses microwave irradiation [39–41] and is further categorized by Sen et al. [40] as stated below:

1. **Microwave Initiated Method:** Using MW irradiation alone for generation of reactive sites.
2. **Microwave-Assisted Method:** Using chemical initiator as well as MW irradiation both for generating reactive sites.

Microwave-initiated method of grafting [41] conforms to the norms of green chemistry, since it reduces the use of chemical-generating free radical sites. Further, the technique is very quick and easy to use (e.g., unlike the *Conventional method* of grafting, no inert atmosphere is required). Another contemporary approach has been the combination of conventional and MW irradiation methods of grafting in an aqueous medium. This approach of grafting is known as *the microwave-assisted method* [4, 5, 42]. The process involves the use of a greener and sustainable alternative such as water in place of any other toxic, organic solvent. Water is the preferred solvent as it is abundant, non-corrosive, and non-flammable in nature. Selectivity provided by microwave irradiation and sustainability provided by the use of water as a solvent make this method a highly reproducible, eco-friendly, and suitable method of grafting [42]. The grafting of polyacrylamide (PAM) chains on a biopolymer (polysaccharide) enhance its intrinsic viscosity and hydrodynamic volume [41], resulting in a more efficient flocculant than the starting material. Figure 9.5

depicts the advantages of using a microwave-based grafting method over a conventional grafting method.

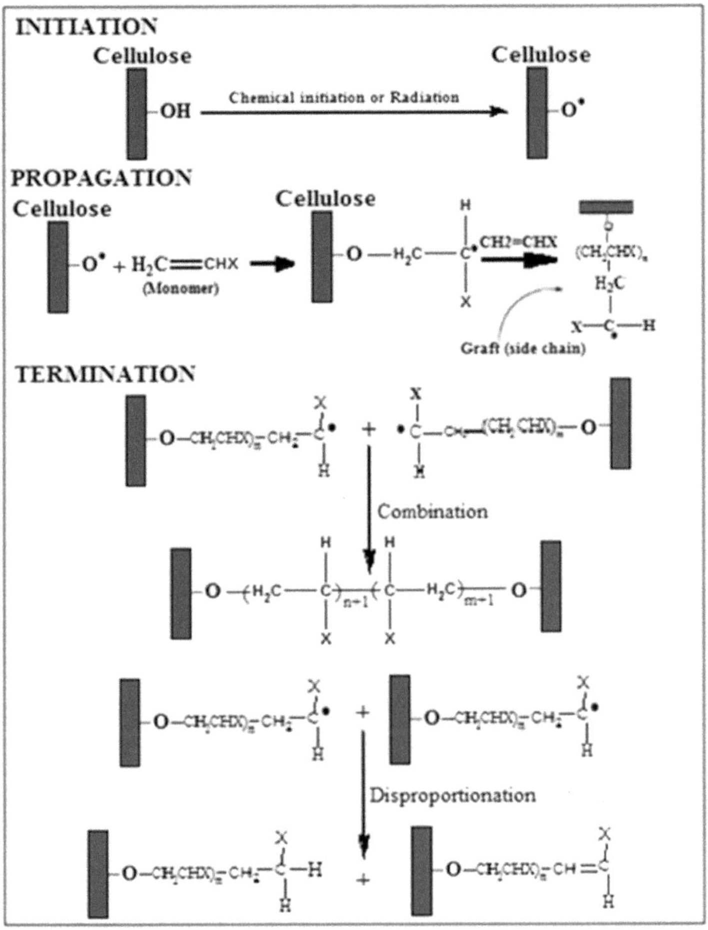

FIGURE 9.5 Initiation of graft polymerization by ceric ion.

9.2.3.2 EVIDENCE OF GRAFTING

Different parameters can be used to confirm polymer modification in terms of grafting. The percentage grafting is the most important characteristic in describing the nature of graft copolymers. Percentage grafting of a graft copolymer is gravimetrically estimated during its synthesis according to the formula mentioned below:

$$\% \text{ Grafting} = [\text{wt. of copolymer} - \text{wt. of polysaccharide}] \times 100 / \text{wt. of polysaccharide} \quad (1)$$

Percentage grafting is a synthesis parameter that profoundly influences the properties of the graft copolymer product. Thus, graft copolymers are in fact tailor-made materials whose properties can be precisely planned at the molecular level (in terms of %grafting) during its synthesis. Higher the extent of grafting (i.e., %grafting), higher is the hydrodynamic volume and molecular weight of the resulting graft copolymers. Consequently, the resulting key alterations in properties for all graft copolymers (in general) are following:

1. **Higher Intrinsic Viscosity:** The intrinsic viscosity of a polymeric solution is exactly proportional to its hydrodynamic volume at the molecular level. The grafting of polymer results in an increased intrinsic viscosity of the graft copolymer resulting in an increase in molecular weight/hydrodynamic volume.
2. **Higher Flocculation Efficacy:** Flocculation efficacy of a molecule is directly proportional to its hydrodynamic volume. It is obvious that graft copolymers having a large hydrodynamic volume have high efficacy as flocculant.

The most commonly used techniques as evidence of grafting are thermal study of mechanical properties, determination of molecular weight, morphological, rheological analysis, and evidence of grafting from spectroscopic tools such as UV-Visible, FTIR, and NMR-spectroscopy.

9.3 RESULTS

9.3.1 APPLICATIONS OF GRAFT COPOLYMER

In recent years, the introduction of graft copolymers has proven to be extremely beneficial. Graft copolymers can be employed as surface modifiers, thermoplastics, medicinal materials, and a variety of other applications [43–47]. The main attraction to graft copolymers, however, is their ability to change the surface of objects. Antistatic agents, surface humidifiers, polymer blend compatibilizers, dye binding intermediates, adhesives, and lubricants are all possible applications for graft copolymers. The drug delivery agent and the flocculant for wastewater treatment are two major uses of graft copolymer generated by the *grafting from* technique. There are many other applications of graft copolymer summarized in Table 9.1.

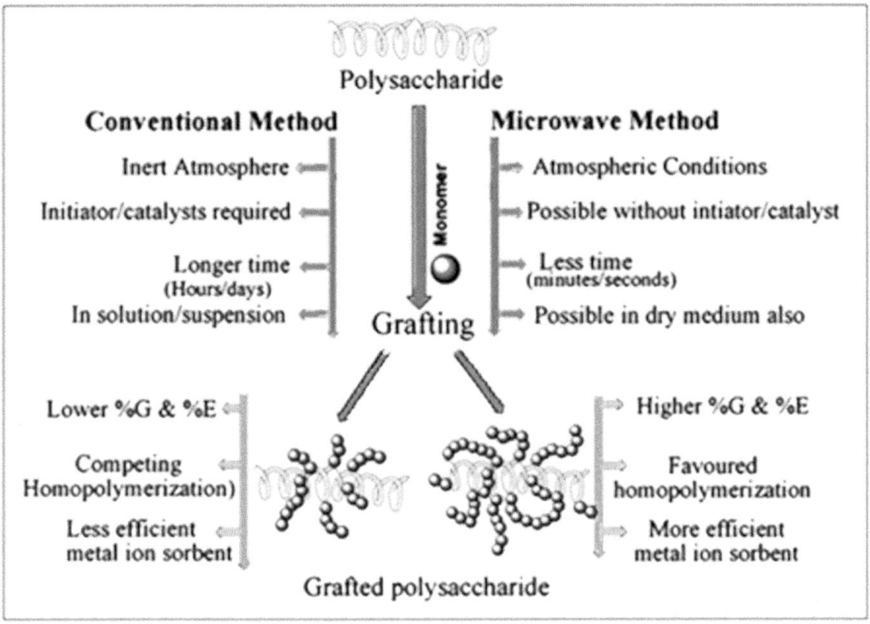

FIGURE 9.6 Schematic diagram represents conventional and microwave induced grafting on polysaccharide backbone.

9.4 CONCLUSION

Graft copolymerization of natural polymers is considered to be a resilient approach, as it functionalizes these biopolymers to their full potential, imparting desirable features. As a result, grafting chemical alterations of natural biopolymers has received a lot of attention. Grafting synthetic polymers onto natural polymers is a simple way to add new features while preserving the substrate's original properties. The grafted copolymer is essential for changing the material's physical-chemical characteristics. Graft copolymerization is a rapid approach to incorporate a desired property into biopolymers. From the above, it is clear that grafted copolymer syntheses are a cost-effective and unique approach with a wide range of applications, including drug delivery, adsorption, textile/tannery wastewater treatment, domestic/sewage wastewater treatment, agriculture, and serving mankind.

TABLE 9.1 Uses and Opportunities of Graft Copolymer

Contemporary commercial applications of graft copolymers:	• Micro porous membranes and nonwoven; • Battery separation membranes; • Release coated film and papers.
Uses in early stages of commercialization:	• Odor absorbent fabrics; • Absorbents for metal ions; • Surface modification of glass windows for ease of cleaning; • Substrates for cell tissue growth.
Developmental uses:	• Fluoropolymer membranes for fuel cells; • Compatibilization in heterogeneous composites.
Possible applications in future:	• Modification of fabrics for flame retardancy; • Biomedical uses as transdermal systems.

ACKNOWLEDGMENT

The author expresses her gratitude to the Birla Institute of Technology, Mesra, Ranchi, Jharkhand, for conducting this study.

KEYWORDS

- **biopolymer**
- **free radical graft copolymerization**
- **grafting**
- **green materials**
- **hybrid materials**
- **microwave-induced techniques**
- **surface modification**

REFERENCES

1. Mohanty, A. K., Misra, & Drzal, L. T., (2002). Sustainable bio-composites from renewable resources: Opportunities and challenges in the green materials world. *J. Polym. and Environ., 10, 19–26.*
2. Wool, R. P., & Sun, S. X., (2005). *Bio-Based Polymers and Composites*. Elsevier Academic Press, Amsterdam; Boston.
3. Sumit, M., Gautam, S., Usha, R. G., & Sweta, S., (2011). Microwave-assisted synthesis of polyacrylamide grafted agar (Ag-g-PAM) and its application as flocculant for wastewater treatment. *International Journal of Biological Macromolecules, 49*, 591–598.
4. Sweta, S., (2018). Derivatization of milk protein using poly (acrylamide): Its characterization and application. *Macromolecular Symposia, 381*(1), 1800120. Wiley.
5. Sweta, S., Gautam, S., & Sumit, M., (2015). In: Sabu, T., Nandakumar, K., Maciej, J., & Josmine, P. J., (eds.), *Advanced Polymeric Materials From Macro-to Nano-Length Scales*. Apple Academic Press, Taylor & Francis Group.
6. Sweta, S., & Gautam, S., (2018). In: Sabu, T., Nandakumar, K., Pious C. V., Zakiah A., & Joseph, T. H., (eds.), *Developments in Preparation, Processing and Characterization of Polymeric and Nano Materials*. Published by Apple Academic Press, Inc. Apple Academic Press, Taylor & Francis Group.
7. Kumar, A., Singh, K., & Ahuja, M., (2009). Xanthan-g-poly(acrylamide): Microwave assisted synthesis, characterization and in vitro release behavior. *Carbohydrate Polymers, 76*, 261–267.
8. Gennadios, A., (2002). *Protein-Based Films and Coatings* (pp. 467–484). *CRC Press, Boca Raton,* FL.
9. Wang, N. G., Zhang, L. N., Lu, Y. S., & Du, Y. M., (2005). Properties of crosslinked casein/waterborne polyurethane composites. *J. Appl. Polym. Sci., 91*, 332.
10. Singh, H., (2006). The milk fat globule membrane: A biophysical system for food applications. *Colloid Interface Sci, 11*, 154–163.
11. Zhang, J., Jiang, L., Zhu, L., Jane, J. L., & Mungara, P., (2006). Morphology and properties of soy protein and polylactide blends. *Biomacromolecules, 7, 1551–15*61.
12. Singh, B., Chauhan, G. S., Kumar, S., & Chauhan, N., (2007). Synthesis, characterization and swelling responses of pH sensitive psyllium and polyacrylamide based hydrogels for the use in drug delivery (I). *Carbohydrate Polymers, 67*, 190–200.
13. Elliott, J. E., Macdonald, M., Nie, J., & Bowman, C. N., (2004). Structure and swelling of poly(acrylic acid) hydrogels: Effect of pH, ionic strength, and dilution on the crosslinked polymer structure. *Polymer, 45*(5), 1503–1510.
14. Sebenik, A., (1998). Living free-radical block copolymerization using this-iniferters. *Prog Polym Sci., 23, 875–917.*
15. Hebeish, A., & Guthrie, J. T., (1981). *The Chemistry and Technology of Cellulosic Copolymers* (p. 351). Springer-Verlag, New York.
16. Moad, G., & Solomon, D. H., (2006). *The Chemistry of Radical Polymerization* (2nd edn.). Elsevier Ltd, Oxford.
17. Rizzardo, D. H., Chiefari, J., Chong, B. Y. K., Ercole, F., Krstina, J., Jeffery, J., Le, T. P. T., et al., (1999). *Macromol. Symp., 143*, 291–307.

18. Moad, G., Rizzardo, E., & Thang, S. H., (2005). Living radical polymerization by the RAFT process. *Aust. J. Chem., 58*, 379–410.
19. Ibrahim, M., Hassan, A., Iqbal, M., & Valeem, E. E., (2008). Response of wheat growth and yield to various levels of compost and organic manure. *Pak. J. Bot., 40*, 2135–2141.
20. Stannett, V. T., & Hopfenberg, H. B., (1971). In: Bikales, N. M., & Segal, L., (eds.), *Cellulose and Cellulose Derivatives* (Vol. 5, pp. 907–936). John Wiley and Sons, New York.
21. Dreyfuss, P., & Quirk, R. P., (1987). *Encyclopaedia of Polymer Science and Engineering* (2nd edn., Vol. 7). John Wiley & Sons.
22. Kobayashi, S., & Uyama, H., (1994). In: Mishra, M. K., (ed.), *Macromolecular Design: Concept and Practice* (Vol. 1, pp. 1–38). Polymer frontiers International Inc., New York.
23. Kunze, J., & Fink, H. P., (2005). Structural changes and activation of cellulose by caustic soda solution with urea. *Macromol. Symp., 223*, 175–187.
24. Andreozzi, L., Castelvetro, V., Ciardelli, G., Corsi, L., Faetti, M., Fatarella, E., & Zulli, F., (2005). Free radical generation upon plasma treatment of cotton fibers and their initiation efficiency in surface-graft-polymerization. *J. Colloid Interface Sci., 289*, 455–465.
25. Ibrahim, M. M., Porreca, F., Lai, J., Albrecht, P. J., Rice, F. L., Khodorova, A., Davar, G., et al., (2005). CB2 cannabinoid receptor activation produces antinociception by stimulating peripheral release of endogenous opioids. *Proc. Natl. Acad. Sci. USA, 102*, 3093–3098.
26. Bhattacharya, A., Rawlins, J. W., & Ray, P., (2008). *Polymer Grafting and Crosslinking.* New York: John Wiley & Sons.
27. Temel, O., & Ismail, C., (2007). Synthesis of block copolymers via redox polymerization process: A critical review. *Iranian Polymer Journal, 16*(8), 561–581.
28. Gupta, K. C., & Khandekar, K., (2006). Ceric (IV) ion induced graft copolymerization of acrylamide and ethyl acrylate onto cellulose. *Polymer International, 55*(2), 139–150.
29. Zahran, M. K., (2006). Grafting of methacrylic acid and other vinyl monomers onto cotton fabric using Ce(IV) ion and cellulose thiocarbonate redox system. *J. Polymer Research, 13*, 65–71.
30. Xi, D., Yang, C., Liu, X., Chen, M., Sun, C., & Xu, Y., (2005). Graft polymerization of styrene on soy protein isolate. *J. Appl. Polym. Sci., 98*(3), 1457–1461.
31. Edimecheva, I. P., Kisel, R. M., Shadyro, O. I., Kazem, K., Murase, H., & Kagiya, T., (2005). Homolytic cleavage of the o-glycoside bond in carbohydrates: A steady-state radiolysis study. *Journal of Radiation Research, 46*, 319–324.
32. Kim, S. J., Lee, C. K., & Kim, S. I., (2004). Electrical/pH-responsive properties of poly(2-acrylamido-2-methylpropane sulfonic acid)/hyaluronic acid hydrogels. *J Appl Pol Sci., 92*, 1731–1736.
33. Fortunati, E., Armentano, I., Zhou, Q., Puglia, D., Terenzi, A., Berglund, L. A., & Kenny, J. M., (2012). Effects of modified cellulose nanocrystals on the barrier and migration properties of PLA nano-biocomposites. *Polymer Degradation and Stability, 97*, 2027–2036.
34. Gubler, L., Slaski, M., Wallasch, F., Wokaun, A., & Scherer, G. G., (2009). Radiation grafted fuel cell membranes based on co-grafting of α-methyl styrene and

methacrylonitrile into a fluoropolymer base film. *Journal of Membrane Science, 339*, 68–77.
35. Sinha, S., Sen, G., & Mishra, S., (2018). Chemical modification of major milk protein (casein): Synthesis, characterization, and application. *Functionalized Engineering Materials and Their Applications*. Publisher CRC Press.
36. Sen, G., & Pal, S., (2009). Microwave initiated synthesis of polyacrylamide grafted carboxymethyl starch (CMS-g-PAM): Application as a novel matrix for sustained drug release, *International Journal of Biological Macromolecules, 45*, 48.
37. Singh, V., Tiwari, A., Tripathi, D. N., & Sanghi, R., (2006). Microwave enhanced synthesis of chitosan-graft-polyacrylamide, *Polymer, 47*, 254–260.
38. Dallinger, D., & Kappe, C. O., (2007). The impact of microwave synthesis on drug discovery. *Chem Rev., 107*, 2563–2591.
39. Sen, G., Sharon, A., & Pal, S., (2011). In: Susheel, K., & Luc, A., (eds.), *Grafted Polysaccharides: Smart Materials of the Future, Their Synthesis and Applications (Chapter 05)*. Hoboken, Wiley-Scrivener.
40. Sinha, S., Mishra, S., & Sen, G., (2013). Microwave initiated synthesis of polyacrylamide grafted casein (CAS-g-PAM)–its application as a flocculant. *International Journal of Biological Macromolecules, 60,* 141–147.
41. Mishra, S., Sinha, S., Dey, K. P., & Sen, G., (2014). Synthesis, characterization and applications of polymethylmethacrylate grafted psyllium as flocculant. *Carbohydrate Polymers, 99*, 462–468.
42. Singh, V., & Tripathi, D. N., (2006). Microwave promoted grafting of acrylonitrile onto *Cassia siamea* seed gum. *J. Appl. Polym. Sci., 101*, 2384–2389.
43. Singh, R. P., Karmakar, G. P., Rath, S. K., Karmakar, N. C., Pandey, S. R., Tripathy, T., Panda, J., et al., (2000). Biodegradable drag-reducing agents and flocculants based on polysaccharides: Materials and applications. *Polym. Eng. Sci., 60*, 40–46.
44. Crini, G., (2005). Recent developments in polysaccharide-based materials used as adsorbents in wastewater treatment. *Prog., Polym. Sci., 31*, 38–70.
45. Singh, B., (2007). Psyllium as therapeutic and drug delivery agent. *Int. J. Pharm., 334*(1, 2), 1–14.
46. Xiang, P., Zhou, T., Wang, L., Sun, C. Y., Hu, J., Zhao, Y. L., & Yang, (2013). *Novel benzothiazole, benzimidazole and benzoxazole derivatives* as potential antitumor agents: Synthesis and preliminary in vitro biological evaluation. *Molecules, 17*(1), 873–883.
47. Sweta, S., (2018). Biodegradable superabsorbents: Methods of preparation and application: A review. *Fundamental Biomaterials: Polymers*, 307–322.
48. Iskandar, S., (2011). *Atom Indonesia, 22*(37), 24–28.

CHAPTER 10

Biodegradable PHAs: Promising "Green" Bioplastics and Possible Ways to Increase Their Availability

TATIANA GR. VOLOVA,[1,2] EVGENIY G. KISELEV,[1,2] ALEKSEY V. DEMIDENKO,[1,2] EKATERINA I. SHISHATSKAYA,[1,2] and SABU THOMAS[1,3]

[1]*Siberian Federal University, 79 Svobodnyi Av., Krasnoyarsk–660041, Russia*

[2]*Institute of Biophysics SB RAS, Federal Research Center "Krasnoyarsk Science Center SB RAS," 50/50 Akademgorodok, Krasnoyarsk–660036, Russia, E-mail: volova45@mail.ru (T. G. Volova)*

[3]*International and Inter-University Center for Nanoscience and Nanotechnology, Mahatma Gandhi University, Kottayam, Kerala, India*

ABSTRACT

Two approaches have been investigated in order to reduce the cost and increase the availability of the use of biodegradable polymers of microbiological origin (polyhydroxyalkanes: PHA): (i) expanding the raw material base and attracting glycerol as a carbon substrate (large-tonnage waste from the production of biodiesel); and (ii) filling the polymer with natural materials for industrial use. An effective technology for the synthesis of destructible polymers was developed using a new and more affordable substrate compared to sugars–glycerol. The *Cupriavidus eutrophus* B-10646 strain was studied as a product of polymers on glycerol, and

Advances in Diverse Applications of Polymer Composites: Synthesis, Application, and Characterization.
Suji Mary Zachariah, Yang Weimin, Maciej Jaroszewski, & Sabu Thomas (Eds.)

the kinetic and production characteristics of the strain providing polymer yields on glycerol of various cheat, comparable with the process on sugars, were studied. Replacing sugars with glycerol provides a reduction in the cost of a carbon substrate in the production of PHA. To increase the availability of PHA and their technical applications, mixtures of PHA with natural materials were formed, and the structure and properties, as well as the patterns of degradation in the soil, were studied. The possibility of long-term functioning of mixed forms in the soil with a gradual release of active substances (herbicides and fungicides) and the possibility of providing plants with protective equipment during the growing season are shown. Both approaches are effective for enhancing accessibility and expanding the scope of application of perspective "green" plastics – destructible PHA.

10.1 INTRODUCTION

The concept of sustainable development is the basic idea of the 21st century. Annual production of synthetic plastics has exceeded 350 million tons, and they largely accumulate in landfills, occupying fertile arable lands. Plastic wastes ruin municipal sewage and drainage systems. Up to 10% of the annual plastic production leaks into the Global Ocean [1]. New approaches to economic management should include the promotion of new functional and environmentally friendly materials, produced without consuming of non-renewable energy resources, while preserving them for future generations. Therefore, the development of new environmentally friendly materials, which will completely degrade without releasing toxic products, should be the priorities for technologists of the 21st century.

Industrial ecology and green chemistry are tools for creating new eco-friendly materials including those manufactured from renewable sources [2]. Hybrid materials based on synthetic polyolefins and natural materials have been increasingly used in various applications. Polymers synthesized by living systems, the so-called biopolymers, have attracted considerable attention, too. Among various natural materials, a special place is occupied by polyhydroxyalkanoates (PHAs) – microbial polymers that can be used in different areas. PHAs are degraded to harmless products (CO_2 and H_2); they are biocompatible, mechanically strong, thermoplastic, capable of blending with other materials, and processable by conventional methods from different phase states [3–6].

To a large extent (up to 45–48%), the cost of PHAs is determined by the cost of high purity carbon substrates [7, 8]; therefore, one of the most high-priority areas of research is the development of technologies using low-cost substrates.

However, the high cost hinders the use of PHAs in technical applications such as agriculture (films and pots for greenhouses, carriers for fertilizers and agrochemicals); production of packages, containers, and household goods; construction industry; etc., which require very large amounts of material. PHA cost can be reduced by upgrading production processes and by using cheaper substrates. Another approach is to use PHA blends with cheaper materials: this can both reduce the cost of the material and modify the properties of the polymers.

There are few published data on the production and properties of PHA blends with solid-phase fillers. Some relatively recent studies investigated natural materials as fillers for modifying properties of PHAs. They described results of filling polyhydroxyalkanoates with clay and its derivatives [9], plant fibers [10], lignin, and holocellulose from a lignocellulosic biowaste [11], and wood chips and powder [12, 13]. As those studies describe different methods of production and investigation of PHA-based blends, it is difficult to compare them and draw any conclusion about the consistent effects of the fillers on the properties of these polymers. So, obtaining composite materials based on PHA is an unconditional resource for increasing their accessibility and expanding the scope of application.

Great potential for reducing the cost of PHA is the use of waste a C-substrate. Potential raw materials for PHA synthesis are various substrates with different oxidation states of carbon, energy content, and cost, including individual compounds (carbon dioxide and hydrogen, sugars, alcohols, organic acids) and byproducts of alcohol and sugar industries, chemical processing of plant raw materials, and production of olive, soybean, and palm oils [14–16]. PHAs can be synthesized from both individual carbon compounds and various industrial wastes, making them economically feasible materials for different applications. Thus, the cost-effectiveness of PHAs can be enhanced by using new strains and carbon sources and by improving manufacturing processes, including large-scale ones. Scaling up laboratory biotechnologies to pilot production (PP) is a necessary step towards commercial production.

One of the promising substrates for large-scale production of PHAs is glycerol, the scale of production of which is currently increasing. This is due to the growing production of biodiesel as an alternative renewable energy source [17]. Biodiesel production increased dramatically from 500,000 gallons in 1999 to 450 million gallons in 2007, and it has been growing steadily over the past few years. In 2010, the global production of biodiesel reached 19 billion L [18], rising to more than 28 billion L in 2017 [19]. Glycerol is a by-product, amounting to about 10%, of biodiesel production by transesterification of animal and vegetable fats and oils (rapeseed, mustard, soybean, and palm oils) [14]. In industrial glycerol grades, the water content varies between 5.3% and 14.2%; methanol content – 0.001–1.7%; NaCl–traces – 5.5%; K_2SO_4 – 0.8–6.6% [17]. The analysis of publications generally indicates that glycerol has the potential to be used as a substrate for PHA production. However, the presence of impurities in glycerol indirectly affects the synthesis of PHA and their properties. It is obvious that involving new strains and upgrading the technological stages of the process will contribute to the improvement of PHA production.

In this work, our results are presented that are aimed at increasing the availability of PHA, which requires the use of glycerol as a substrate [20–23], as well as the preparation of PHA composites with available natural materials [24, 25].

10.2 MATERIAL AND METHODS

The strain used in this study was *Cupriavidus eutrophus* B-10646, registered in the Russian Collection of Industrial Microorganisms (RCIM). Schlegel's mineral medium was used as a basic solution for growing cells: Na_2HPO_4-H_2O–9.1; KH_2PO_4–1.5; $MgSO_4$-H_2O–0.2; $Fe_3C_6H_5O_7$-$7H_2O$–0.025; $CO(NH_2)_2$–1.0 (g/L). Nitrogen was provided in the form of urea, and, thus, no pH adjustment was needed. The pH level of the culture medium was stabilized at 7.0±0.1 To cultivate bacteria in shake flask culture, an Innova 44 constant temperature incubator shaker (New Brunswick Scientific, U.S.) was used. Inoculum was prepared by resuspending the reference bacterial culture maintained on agar medium. The reference culture was grown in 1.0–2.0 L glass flasks half-filled with liquid saline medium, with the initial concentration of glycerol from 5 to 10 g/L. Growth kinetics of bacterial cells was studied in automated laboratory fermenters (Bioengineering AG, Switzerland), with 30-L and

150 L fermentation vessels and the working volume of the culture from 18 to 100 L, under strictly aseptic conditions [21].

Fermenters were equipped with systems for monitoring pH level, foam level, temperature, pressure, and dissolved oxygen. The fermenters were controlled by BioScadaLab software in automatic mode. To supply the feeding substrates, the fermenters were equipped with Bioengineering Peripex peristaltic pumps. The concentration of dissolved oxygen was maintained at DO 30%. The air supply control was carried out in a cascade mode (DO-air flow-mixer revolutions). During cultivation in a 30-L fermenter, the amount of air supplied per cultivation process varied from 0 to 5.5 Nl/min, the speed varied from 500 to 1,000 rpm. During cultivation in a 150-L fermenter, the amount of air supplied per cultivation process varied from 10 to 200 Nl/min, the speed varied from 300 to 750 rpm.

A two-stage process was used. In the first stage, cells were grown under nitrogen deficiency: the amount of nitrogen supplied in this stage was 60 mg/g cell biomass synthesized (i.e., 50% of the cell's physiological requirements – 120 mg/g); the cells were cultured in complete mineral medium and with glycerol flux regulated in accordance with the requirements of the cells. In the second stage, cells were cultured in the nitrogen-free medium; the other parameters were the same as in the first stage. The temperature of the culture medium was 30±0.5°C and pH was 7.0±0.1 [22].

During the cultivation, samples of culture medium were taken for analysis every 4–5 h (from fermenters) or every 8–10 h (from flasks); cell concentration in the culture medium was determined based on the weight of the cell samples dried at 105°C for 24 h (CDM); the total biomass (X_{total}) and catalytically active biomass (X_c) ($X_c = X_{total}$–PHA), g/L, were distinguished. Cell concentration in the culture medium was determined every hour by converting the optical density of culture broth at 440 nm to cell dry mass by using a standard curve prepared previously. The criteria for evaluating the process of PHA biosynthesis were as follows: concentration of cell biomass in culture, polymer content in cells, consumption of the main growth substrate, duration, and productivity of the process. The kinetic and production parameters of the culture were determined by conventional methods. The biomass concentration (X, g/L), the catalytic biomass concentration (X_c, g/L), the yield coefficient of the polymer (Y, g PHA/g substrate), the specific growth rate (μ, h^{-1}), and the volumetric productivity (P, g/(L·h)) were calculated [26].

Intracellular polymer content at different time points was determined by analyzing samples of dry cell biomass. Intracellular PHA content and composition of extracted polymer samples were analyzed using a GC-MS (6890/5975C, Agilent Technologies, U.S.).

Three natural materials were used as fillers: peat, wood flour and clay. High-moor peat "Agrobalt-N," state registration 0428-06-209-139-0-0-01, was produced by OOO "Akademiya Tsvetovodstva," Russia) [25]. Wood flour was produced by grinding wood of birch (*Betula pendula* Roth) using an MD 250-85 woodworking machine ("StankoPremyer" Russia). Then it was dried at 60°C for 120 h until it reached constant weight, and 0.5 mm mesh was used to separate the particle size fraction. Clay was taken from the mine "Kuznetsovskoye," the Krasnoyarskii Krai, Russia. Its composition: loss on ignition–6.92%; SiO_2–60.1%; Al_2O_3–19.17%; Fe_2O_3–6.72%; CaO–2.02%; MgO–2.12%; SO_3–0.65%; Na_2O–0.88%; K_2O–1.45% [24, 25].

The polymer [poly-3-hydroxybutyrate] and fillers were pulverized by impact and shearing action in ultra-centrifugal mill ZM 200 (Retsch, Germany). To achieve high fineness of polymer grinding, the material and the mill housing with the grinding tools were preliminarily cooled at –80°C for about 30 min in an Innova U101 freezer (NEW BRUNSWICK SCIENTIFIC, U.S.). Grinding was performed using a sieve with 2-mm holes at a rotor speed of 18,000 rpm. Fillers were not precooled before grinding; a sieve with 2-mm holes was used, and the grinding was performed at 6,000 rpm. Fractionation of the polymer and filler powders was carried out using vibratory sieve shaker AS 200 control (Retsch, Germany). Fractioning time was 10 min, amplitude 1.5 mm; a 100 g sample was used. Fractions of the polymer and filler powders of 200 μm and less were selected and used. Then, polymer powder was mixed with different filler powders in benchtop planetary mixer SpeedMixer DAC 250 SP (Hauschild Eng., Germany); the blend time was 1 min, and the speed was 1,000 rpm. The blends contained different amounts of the fillers: 10, 30, and 50%.

The homogenized blends were used to produce pellets and granules. Pellets were prepared by cold pressing at 36 bar using a laboratory semi-automatic press (Minipress, Minsk, Belarus). Granules were prepared by mixing wet polymer paste (40%) in ethanol in planetary mixer SpeedMixer DAC 250 SP (Hauschild Eng., Germany), with the mixing time 1 min at 1,000 rpm, and granulating in screw granulator Fimar EAC (Italy) with a

6 mm nozzle. The granules were dried in a fume hood at room temperature for 24 h.

Poly-3-hydroxybutyrate, peat, clay, and wood powders, powdered P(3HB)/filler blends, and granules and pellets were examined by using state-of-the-art physicochemical methods (SEM microscopy, FTIR spectroscopy, differential scanning calorimetry (DSC), X-ray diffraction (XRD); described in detail [26].

10.3 RESULTS AND DISCUSSION

10.3.1 SYNTHESIS OF PHA ON GLYCEROL OF VARIOUS PURIFICATION

Three brands of glycerin were used: Glycerol purified (Corporate Oleon, Sweden): glycerol–99.3; chloride–0.0001; salts (NH_4)–0.005; Fe–0.0005; Ar–0.00004; moisture–0.09; fatty acid and ester–0.25 (% mass); max heavy metal–0.00005 μg/g (Glycerol I). Glycerol refinery B.V. (Dutch glycerol refinery, Netherlands): glycerol–99.7; chloride <0.001; moisture–0.09; fatty acid and ester–1.0; sulfate <0.002; total organic impurities–0.5–1.0; individual organic impurities–0.1 (% mass); heavy metal <5 (μg/g) (Glycerol II). Crude glycerol (Prisma Commercial Exportadora de Oleoquimicos LTDA, Brazil): glycerol–82.1; chloride–4.35; mongo–0.13; methanol–0.13; ash–6.59; moisture–9.88 (% mass); and pH–5.8 (Glycerol III).

10.3.1.1 THE EFFECT OF GLYCEROL CONCENTRATION ON BACTERIAL GROWTH AND PHA SYNTHESIS

C. eutrophus B-10646 cells were cultivated in the media with concentrations of purified glycerol varied within a wide range to determine the limits of physiological action of this substrate for the strain and the kinetic constants [21, 22]. The limits of physiological action of glycerol for this strain are very wide, varying between 0.5 g/L and 60.0 g/L. There was a wide plateau (from 1 to 30 g/L), and the zones of limitation and inhibition of bacterial growth by glycerol were 0.1–3.0 and 30–60 g/L, respectively. The dependence of the specific growth rate (μ) on the substrate concentration (S) was described by the Andrews equation, which is a modified Monod equation.

Using the graphical analysis method of Lineweaver-Burke (1/μ:1/S) and Dickson's method (1/μ:S), we calculated kinetic constants for this strain (saturation constant (K_S) and inhibition constant (Ki), μ_{max}). For the strain *C. eutrophus* B-10646, the limits of the physiological action of glycerol were found to be 1–30 g/L; K_S and K_i were 0.36 g/L (0.004 mol/L) and 62.0 g/L (0.673 mol/L), respectively; $\mu_{max} = 0.085\ h^{-1}$.

10.3.1.2 A STUDY OF THE GROWTH AND SYNTHESIS OF PHA BY C. EUTROPHUS B-10646 ON PURIFIED AND CRUDE GLYCEROL

The content of the main component (glycerol) in purified glycerol is more than 95–99%. In unpurified (crude) glycerol, depending on the raw material and the technology used, the content of glycerol is 80–85%; the other part comprises impurities, including free fatty acids (FFA) and methyl esters of FFA, alcohols, as well as water and salts, which, as a rule, inhibit the microorganisms responsible for PHA production.

The productivity of culture of *C. eutrophus* B-10646 on different glycerol is shown in Table 10.1 [21, 22]. Two grades of purified glycerol, with a 99.3 (Glycerol I) and 99.7% (Glycerol II) content of the basic material, and unpurified, crude glycerol (82.1%) were used in this study. Cultivation of *C. eutrophus* B-10646 cells was carried out in a 30-L fermenter with a starting cell concentration in the inoculate of 1.0–1.5 g/L, 25 g/L and 18 L working volume of the culture. The process was performed in a two-stage culture (30 hours) on saline medium: the growth of cells was limited by nitrogen deficiency in the first stage for 30–32 h; in the second stage (24–30 h), cells were grown in the nitrogen-free medium. The concentration of glycerol in the culture was maintained at 5–10 g/L, and it was fed into the culture with a peristaltic pump dispenser.

TABLE 10.1 Production Indicators of PHA Biosynthesis in a Culture of *C. eutrophus* B-10646 on Glycerol of Various Purification

Type of Glycerol	X (g/L)	PHA (%)	μ* (ч⁻¹)	Y_x	Y_p	Productivity (P) (g/L*h)	
						P_x	P_p
Glycerol-I	69.3 ± 3.5	72.1 ± 3.6	0,101	0.38	0.29	1.10	0.81
Glycerol-II	69.8 ± 3.5	73.3 ± 3.6	0,097	0.39	0.29	1.15	0.90
Glycerol-III	69.3 ± 2.9	73.3 ± 3.2	0,100	0.36	0.26	1.15	0.89

Comparable cell concentrations and polymer contents were achieved in the experiments with *C. eutrophus* B-10646 cultivated on purified glycerol (Glycerol I and Glycerol II) (Table 10.1). When glycerol purified and glycerol refinery B.V. were used, the maximum yield of total biomass was 70 g/L, and the polymer content in cells was 72–75%. The analysis of the results revealed some differences in the kinetic parameters during the cultivation of bacterial cells. As shown in Figure 10.3, the specific growth rate of the total and active cell biomass and consumption of purified Glycerol I varied during the experiment. The specific growth rate of cells cultivated on glycerol purified was the highest (estimated from total and active biomass) in the initial period of the first stage of cultivation on complete nutrient medium with a limited supply of nitrogen (50% of the physiological requirement of bacteria), reaching 0.15 h^{-1} and 0.14 h^{-1}, respectively (Figure 10.3). This period corresponded to the most active consumption of glycerol by the culture, at an average specific rate of 4.0±0.2 g/(g·h). The rate of polymer synthesis at this stage was 0.18 h^{-1}, and tended to decrease. After 30–32 hours, the total cell biomass concentration was 42.1±1.7 g/L, and the polymer content in the cells reached 47.8±2.3%. During the second stage, the supply of nitrogen to the culture was stopped; the controlled supply of glycerol and mineral elements precluded the deficiency of these substrates in the culture. In the second stage, the specific growth rates of bacterial cells and polymer synthesis decreased gradually, as the consumption of glycerol dropped. At the end of the fermentation period (60 h), the total cell concentration was 69.3±3.5 g/L, and the polymer content in the cells was 72.4±3.6%. The consumption of glycerol for the whole period was 3.5±0.2 kg.

On the second type of purified glycerol (Glycerol II), the maximum values of the specific growth rate of bacteria (in terms of total and active biomass) in the first stage of the process were 0.15 h^{-1} and 0.13 h^{-1}, and they were comparable to the results for glycerol purified during this period. The consumption of glycerol in this period was between 3.8 and 4.0 g/(g·h) while the polymer synthesis rate was 0.18 ± 0.02 h^{-1} and tended to decrease. After 30 hours, the total cell biomass concentration was 45.6±2.2 g/L, and the polymer content in the cells reached 56.1±2.7%. At the end of the fermentation period, the total cell biomass concentration was 69.4±3.5 g/L, and the polymer content in the cells was 73.3±3.6%. The consumption of glycerol for the whole period was 3.4±0.2 kg (Table 10.1).

On glycerol III, the maximum values of bacteria specific growth rate (in terms of total and active biomass) in the first stage of the process were 0.14 h^{-1} and 0.13 h^{-1}, respectively, when glycerol was consumed by the culture at a rate of 4.2±0.2 g/g h. The average rate of polymer synthesis in this stage was 0.17±0.02 h^{-1}. After 30 hours, the total cell biomass concentration was 46.2±1.9 g/L, and the polymer content in the cells reached 52.2±2.1%. At the end of the fermentation period, the total cell biomass concentration was 69.3±2.9 g/L, and the polymer content was 78.1±3.2% (Table 10.1). The consumption of raw glycerol for the whole period amounted to 3.8±0.2 kg, which corresponded to the yield coefficient for the polymer – 0.26±0.01, taking into account that the concentration of raw glycerol was 82.07%.

The consumption of raw glycerin for the entire period was 3.8 ± 0.2 kg, which corresponds to the economic coefficient for the polymer Yp – 0.26 ± 0.02. Considering that its concentration in crude (raw) glycerin is 82.07%, the economic coefficient in terms of absolute glycerin is comparable to the values obtained with purified glycerin (Yp 0.29 ± 0.02 g / g).

Thus, it was shown that the production characteristics of *C. eutrophus* B-10646 when using glycerol of various purifications are comparable; the use of crude glycerol did not inhibit bacterial growth and PHA synthesis.

10.3.1.3 PILOT PRODUCTION OF PHA ON GLYCEROL

The process of PHA synthesis by *C. eutrophus* B-10646 on Glycerol II was scaled-up and studied under pilot production (PP) conditions [22]. Pilot production consisted of units for media and inoculum preparation; a unit for fermentation; a unit for polymer extraction and purification. The PP fermentation unit included a steam generator (Biotron, South Korea) for sterilizing fermenters and service lines, a compressor (Remeza, Belarus) for air supply, a 30-L seed culture fermenter, a 150-L production fermenter, an ultrafiltration unit (Vladisart, Russia) to concentrate the culture, and a unit for cool dehumidification of the condensed cell suspension (LP10R ILSHIN C, South Korea).

The *C. eutrophus* B-10646 seed culture was derived from a reference stock culture maintained on agar medium, by growing cells in 2.0-L flasks on a complete nutrient medium in an incubator shaker. The resulting culture was concentrated by centrifugation in compliance with the sterility rules. Inoculum (4 L) with a cell concentration of 14–17 g/L was inoculated into a 30-L fermenter containing phosphate buffer. The process of building up the seed material was carried out on complete nutrient medium. For this purpose, dosing pumps were used to feed continuously separate flows of the solutions of glycerol, urea, magnesium sulfate, and ferric citrate with trace elements into the fermenter. The residual concentration of glycerol in the culture was maintained at 5–20 g/L; urea – 0.1–0.2 g/L; magnesium sulfate – 0.05–0.1 g/L.

The inoculum produced in a seed fermenter, was pumped into a 150-L production fermenter containing a sterile buffer solution of potassium and sodium phosphates in a sterile seeding line; thus, the initial concentration of the culture was 7.0 g/L. The process was carried out in two stages with continuous supply of sterile air and feeding solutions. At the end of fermentation (45 h), the cell biomass concentration was 110±5.5 g/L, and the polymer content 78±3.1%. Thus, the average productivities of the polymer and cell biomass were 1.83 and 2.29 g/(L·h), respectively.

The results of the study of the chemical composition and physicochemical properties of PHA samples synthesized on glycerol of different purification degrees are presented in Table 10.2.

TABLE 10.2 Physicochemical Properties of Samples P(3HB)] Synthesized on Glycerol of Various Purities

Substrate	M_w (кDa)	D	C_x (%)	T_g (°C)	T_c (°C)	T_{melt} (°C)	T_{degr} (°C)
Glucose	920	2.52	76	–	92	178	295
Glycerol-I	355	3.42	50	2.9	96	174	296
Glycerol-II	416	3.63	55	–	103	176	296
Glycerol-III	304	3.49	52	2.7	99	172	295

Note: "–": means no peak.

The properties of PHA samples synthesized by *C. eutrophus* B-10646 on three types of glycerol did not differ dramatically. The polymers synthesized by the glycerol-adapted productive culture *C. eutrophus* B-10646 on two types of purified glycerol had similar values of M_n – 104 and 115 kDa, M_W – 355 and 416 kDa, and polydispersity – 3.42 and 3.63, respectively. The M_n and M_W values of the PHAs produced on crude glycerol were somewhat lower, 87 and 304 kDa. These values were generally lower than those obtained earlier on other substrates. The ^{1}H-NMR spectra of three PHA samples synthesized on three glycerol grades were similar (Figure

10.5) and showed the expected resonances for P(3HB) as demonstrated by the methyl group at 1.25 ppm, the methylene group between 2.45 ppm and 2.65 ppm, and the methine group at 5.25 ppm. The ^{1}H-NMR-obtained spectra of PHA samples synthesized on glycerol were similar to those obtained by other researchers [27–29]. The P(3HB) samples synthesized on glycerol had a reduced C_X (50–55%), i.e., the amorphous and crystalline regions had become nearly equal to each other. No deviations in the temperature properties were found in the samples of P(3HB). T_m and T_d values were within the previously identified value limits, 172–176 and 295–296°C, respectively.

It should be noted that the specific consumption of various substrates (sugars and glycerol) is comparable for the synthesis of PHA by the strain *C. eutrophus* B-10646. Table 10.3 [22] presents a comparison of unit costs and the cost of various carbon substrates for the production of PHA.

It is shown that fructose accounts for the largest costs and amounts to 303.0 rubles (₧)/ kg of polymer. When replacing fructose with glycerin, the cost of the carbon substrate is reduced by 41.6% when using purified glycerin and by 55% when using "raw" glycerin. When glucose is replaced with glycerin, the cost of the carbon substrate is reduced by 8.4% when using purified glycerol and by 29.4% when using "raw" glycerin. Thus, the technology for the synthesis of PHA on glycerol, developed, and implemented in a pilot version, provides an efficient process for producing polymers with a significant reduction in production costs.

TABLE 10.3 Costs of Carbon Substrates for the Production of 1 kg of PHA

Name of substrate	Cost of substrates, ₧/kg	Substrate consumption, kg/kg PHA	The unit cost of substrate, ₧/kg PHA
Fructose	101.0	3.0	303.0
Glucose	69.0	2.8	193.2
Purified Glycerol "Dutch Glycerol Refinery," Netherlands	59.0	3.0	177.0
Crude Glycerol "PRISMA Commercial exporter of oleochemicals LTDA," Brazil	44.0	3.1	136.4

Source: Reprinted with permission from Ref. [22]. © 2019 American Chemical Society.

10.3.2 COMPOSITES BASED ON PHA AND NATURAL MATERIALS

10.3.2.1 CHARACTERIZATION OF P(3HB) AND INITIAL NATURAL MATERIALS

Poly-3-hydroxybutyrate and natural materials (clay, peat, and birch wood flour) with considerably different properties (Table 10.4) were used to produce P(3HB)/filler blends: fine powder and paste prepared with ethanol (40% moisture content) [25].

TABLE 10.4 Physicochemical Properties of Initial Materials and Blends

Sample	C_x (%)	$T_{melt.}$ (°C)	$T_{cryst.}$ (°C)	$T_{degr.}$ (°C)	Melting Enthalpy (J/g)
		Initial Materials:			
P(3HB)	75	176	108	287	89.3
Clay	53	–	–	200*	–
Peat	9	–	–	130*	–
Wood flour	26	–	–	220*	–
		Blends (P(3HB)/Filler = 70/30):			
P(3HB)/Clay:					
90:10	72	177	110	286	52.4
90:30	60	175	110	283	38.5
90:50	56	176	109	276	29.8
P(3HB)/Peat:					
90:10	60	174	107	288	64.1
90:30	48	176	108	292	59.7
90:50	42	177	110	290	48.7
P(3HB)/Wood Flour:					
90:10	65	178	107	279	66.9
90:30	47	175	106	273	51.3
90:50	44	171	104	274	45.4

Note: *: start of $T_{degr.}$

Source: Reprinted with permission from Ref. [25]. © 2020 Elsevier.

P(3HB) is a highly crystalline material, with the prevailing crystalline phase, and its degree of crystallinity (C_X) is usually above 60–65%;

the C_X of the P(3HB) sample used in this study is 75%. XRD pattern of P(3HB) is showed, that diffraction peaks in 2θ = 13.4, 16.8, 20, 22.2, and 25.5°. XRD patterns of the fillers differed from the diffraction pattern of P(3HB). Wood flour showed distinct diffraction peaks at 2θ = 14.2, 19.6, and 25.5°, in contrast to the amorphized peat. X-ray examination of clay revealed numerous narrow diffraction peaks, and the most distinct one was at 23.5°. Clay had the highest C_X (53%). In wood flour and, especially, peat, amorphous phase prevailed, and their degrees of crystallinity were low: 26 and 9%, respectively.

Thermal properties of the polymer and fillers are different too. P(3HB) is a thermoplastic material, with a considerable difference between melting point (176°C) and thermal decomposition temperature (287°C). Its peak of thermal decomposition is in the 280°C. P(3HB) crystallization temperature is 108°C and melting enthalpy is 89.3 J/g. By contrast, none of the fillers is a thermoplastic material. Initial heating of these materials (and blends) gives a small peak at a temperature between 70°C and 130°C, which is associated with evaporation of moisture and volatile organic compounds. After moisture evaporation, thermal decomposition of peat begins at 130°C and thermal decomposition of wood flour at 220°C. For clay, at a temperature between 60°C and 200°C, moisture evaporates and decomposition and evaporation of volatile organic compounds begin (the process is endothermal); however, at temperatures between 200°C and 400°C, decomposition is accompanied by heat release.

Possible structural differences between the blends were detected using IR spectroscopy, which is employed to study the structures of various macromolecules. Variations in the intensity of the bands in the low-frequency range are indicators of both the crystalline to amorphous phase ratio in the blend and the type of interactions of the components in the blend.

FTIR spectra of P(3HB) and the three fillers were taken in the 400–4,000 cm^{-1} range. The spectra show not only the groups of carbon compounds (C=O, C-OH, CH_2, CH_3) contained in all initial materials but also the groups and compounds characteristic of individual components. These are CH, CH_2, CH_3, C-OH, and COOH groups for P(3HB), S-S-, C=C, C=N, and N–H groups for peat, Si-O-Si(AL) and Fe-O in α-Fe_2O_3 for clay, and -C-O-;-C-O-C- for birch wood flour. The IR absorption spectra of P(3HB) (Figure 10.3) contain absorption bands corresponding to vibrations of the main structural components of polymers except the absorption bands of vibrations of the terminal C-OH and COOH groups. The bands of the ordered optical densities (crystalline phase) are in the 1,228 cm^{-1} range while the bands of the amorphous phase are shifted to 1,182 cm^{-1}. The spectra show distinct absorption bands of asymmetric stretching of CH_3- and CH_2-groups (2,978 and 2,994 cm^{-1}); symmetric stretching of CH- and CH_2 groups (2,994 and 2,934 cm^{-1}); stretching of conjugated (1,687 cm^{-1}) and unconjugated (1,720 cm^{-1}) carbonyl groups C=O; skeletal CH vibrations (599 cm^{-1}); and CH bending (622 cm^{-1}). PHAs are hydrophobic compounds, in which molecules of water interact with the hydrogen atom of the methyl group or with the oxygen atom of the carboxyl group through hydrogen bonding. The narrow bands at 3,436 cm^{-1} are typical of the hydrogen-bonded OH group. The 1,687 cm^{-1} band is characteristic of hydrogen vibration in the OH group interacting with the oxygen atom through hydrogen bonding. The 2,874, 2,934, 2,978, and 2,994 cm^{-1} absorption bands are typical of the bonded OH group and may be a component of the COOH dimer group. In the 3,000–2,500 cm^{-1} range, there is a group of weak bands typical of the dimers of carboxylic acids.

In contrast to P(3HB) IR spectra, the IR spectra of peat contain absorption bands of the -S-S-bond groups (521 cm^{-1}); -C-O-groups characteristic of carbohydrates (1,029 cm^{-1}); deformation vibrations of -CH_2-groups (718 cm^{-1}); skeletal vibrations of the benzene ring C=C (lignin), C=N, N–H (1,513 and 1,597 cm^{-1}), and stretching of the –OH groups at 3,300 cm^{-1}. The IR spectra of clay contain absorption bands in the 776 and 795 cm^{-1} range, corresponding to stretching and symmetrical vibrations of Si-O-Si(Al) and Fe-O in α-Fe_2O_3; stretching vibrations of (AI-0), (Si-O-Al), and -$A1_2O_3$ in the 692 cm^{-1} range. In the 978 cm^{-1} range, there are absorption bands corresponding to (O-Si) deformation vibrations. The absorption frequency observed at 1,635 cm^{-1} is indicative of deformation vibrations of H-O-H, which suggest water sorption on the surface of the mineral. The absorption band at 3,615 and 3,392 cm^{-1} corresponds to stretching and symmetric vibrations of AlO-H for clay bound water. IR spectra of birch wood flour show plane bending of the guaiacyl ring of lignin of leaves (1,031 cm^{-1}), stretching of -C-O-; -C-O-C- (1,235 cm^{-1}), skeletal vibrations of the syringyl ring of lignin of leaves (1,324 cm^{-1}), stretching, and skeletal vibrations of benzene ring (1,422 cm^{-1}, 1,504 cm^{-1}, 1,593 cm^{-1}), stretching of -C=O of conjugated and unconjugated groups (1,652 cm^{-1}, 1,734 cm^{-1}), and intramolecular and intermolecular stretching of –OH groups and water (3,337–3,342 cm^{-1}).

10.3.2.2 CHARACTERIZATION OF P(3HB)/FILLER BLENDS

Powdered P(3HB) and fillers were processed by cold pressing to produce pellets with a diameter of 6 mm and height 3 mm and granules with a diameter of 2 mm and length of between 2 mm and 4 mm (Figure 10.1).

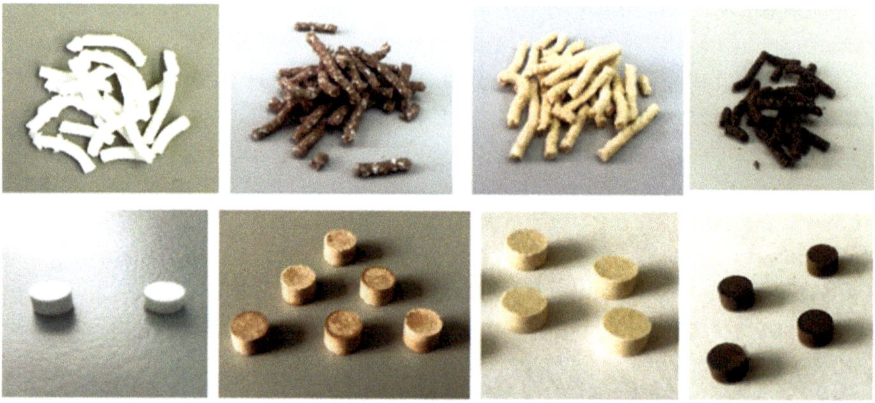

FIGURE 10.1 Granules and 3D forms obtained from PHA mixed with natural materials.
Source: Reprinted with permission from Ref. [23]. © 2019 Elsevier.

Pellets and granules prepared from blends with fillers that had different properties exhibited diverse surface structures (Figure 10.2). SEM images [25] showed porous and rough surface microstructures. The surface of the P(3HB)/wood flour pellets showed rectangular sawdust particles embedded in the surface layer; on the surface of the P(3HB)/peat pellets, there were asymmetrical heteraxial inclusions. On P(3HB)/clay pellets, clay inclusions were uniformly distributed over the surface.

The blends were characterized using DSC, X-ray examination, and IR spectroscopy. IR spectrometry showed that characteristic absorption bands of initial components were separately retained in all blends: for P(3HB) those were CH, CH_2, CH_3, C-OH, and COOH groups; for peat–the additional and characteristic S-S-, C=C, C=N, N–H groups; for clay–Si-O-Si(AL) and Fe-O in α-Fe_2O_3; for birch wood flour – -C-O-;-C-O-C-. In the spectra of the blends, absorption bands were either separate bands or coinciding peaks, some of them transformed. An insignificant difference in numbers was within the limits of spectral error – 4 cm^{-1} (accumulation of 256 scans) and could result from increased moisture content of some of the samples. No new significant absorption bands were revealed in the blends, suggesting that no new chemical bonds were formed and that the blends prepared by the techniques employed in this study were physical mixtures.

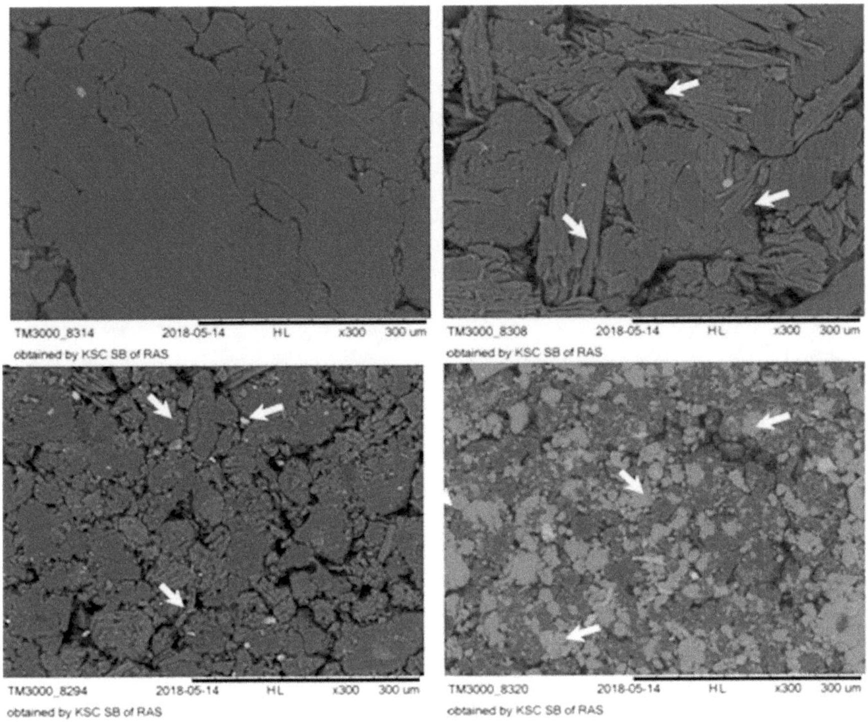

FIGURE 10.2 SEM images of pressed pellets (P(3HB)/filler = 70/30): (a) P(3HB); (b) P(3HB)/wood flour; (c) P(3HB)/peat; (d) P(3HB)/clay (arrows–turning on fillers).

Source: Reprinted with permission from Ref. [25]. © 2020 Elsevier.

Schematically, the process of obtaining mixtures from polymers and natural materials is presented in Figure 10.3.

Blending of P(3HB) with the fillers changed the properties of the polymer. The blends became progressively amorphous as the percentage of the fillers was increased (10, 30, and 50% fillings of the polymer with natural materials were tested). The degree of crystallinity of P(3HB) was decreased more noticeably in the blends with peat and wood flour (materials with low C_X). The C_X of the P(3HB)/peat and P(3HB)/wood flour blends with 30% filler decreased to 47 and 48%, respectively; the

C_X of the blends with 50% of the same fillers decreased to 42 and 44%, respectively. That is, amorphous phase prevailed in these blends. In the P(3HB)/clay blends, the crystalline phase prevailed at all levels of filling of the polymer with clay; the C_X was no less than 56%. Changes in the crystallization kinetics of the polymer in the blends were associated with the rather high percentages of fillers (30 and 50%). This must-have decreased the free volume necessary for nucleation of spherulites to occur, preventing the development of the crystalline phase and, thus, reducing it.

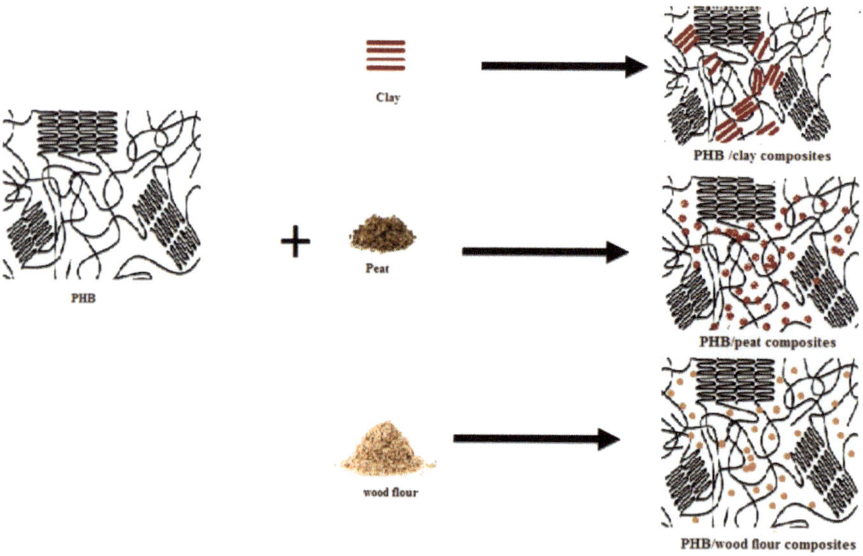

FIGURE 10.3 Schemes of initial materials and formulated mixtures [25].
Source: Reprinted with permission from Ref. [25]. © 2020 Elsevier.

The study of thermal properties of the blends did not show any significant influence of the fillers on the temperatures of melting and thermal decomposition and the position of the T_{melt} and T_{degr} peaks in thermograms. However, the onset of thermal decomposition of the blends was shifted to the left relative to the initial P(3HB). The enthalpy of melting of the blends was significantly decreased (by a factor of 1.5–2.2). It is difficult to compare results obtained in this study with literature data because of data scarcity and different techniques used to process blends into products.

The study of physicochemical properties of P(3HB) blends with the three natural materials did not reveal any dramatic impairment of the properties of the polymer. However, Young's moduli of all blends decreased as the amount of the filler in the blend increased. The most significant decrease in Young's modulus was observed in the P(3HB)/peat and P(3HB)/clay blends containing 50% of the filler: the parameter dropped to 150 and 76 MPa, respectively, compared to Young's modulus of the initial polymer of approximately 360 MPa.

As P(3HB) blends can be considered as potential carriers for pesticides and fertilizers for soil application and as materials for fabricating degradable packaging and other goods, it is important to study their degradation behavior in soil. Degradation of P(3HB) blends with natural materials was first investigated in this study, showing that the mass loss of the blends over 35 days of incubation in soil varied between 30% and 50% of the initial mass of the products, depending on the type of the filler [25].

In this way, natural materials–clay, birch wood flour, and peat–were used as fillers for degradable poly-3-hydroxybutyrate. The P(3HB)/filler to produce of pellets and granules. All initial materials and blends were investigated using IR spectroscopy, DSC, X-ray analysis, electron microscopy with microanalysis and X-ray spectrometry. Analysis of IR spectra showed that no chemical bonds were established between the polymer and fillers and that the blends were physical mixtures. Their temperature characteristics and degrees of crystallinity were lower than those of P(3HB), suggesting different crystallization kinetics of the blends. Degradation behavior of pellets and granules prepared from P(3HB)/filler blends in soil was similar to the degradation behavior of pure polymer. The mass loss of all pellets and granules over 35 days of incubation in soil varied between 30% and 50% of the initial mass of the products, depending on the type of the filler. Thus, the blends prepared and tested in this study can be used to fabricate degradable packaging, greenhouse accessories, and delivery systems for fertilizers and pesticides.

10.3.3 A STUDY OF THE SUITABILITY OF MIXTURES OF P(3HB) AND NATURAL MATERIALS FOR THE CONSTRUCTION OF PLANT PROTECTION PRODUCTS OF A NEW GENERATION

The newest trend in research is the development of new-generation slow-release pesticides, including fungicides, embedded in degradable matrices. The use of such formulations will decrease the amounts of pesticides applied to soil, ensure slow and targeted release of the active ingredients, and reduce the risk of uncontrolled distribution of the pesticides in the biosphere. The main condition for constructing such formulations is the availability of appropriate materials with the following properties: degradability and compatibility with global biosphere cycles, safety for living organisms and their nonliving environment, long-term presence in the natural environment and controlled degradation followed by formation of non-toxic products, chemical compatibility with pesticides, and processability by available methods.

Therefore, the purpose of this study was to construct and investigate slow-release fungicide formulations, with the active ingredient embedded in the degradable matrix of P(3HB) blended with readily available natural materials (peat, clay, wood flour) [23]. Slow-release fungicide formulations (azoxystrobin, epoxiconazole, and tebuconazole) shaped as pellets and granules in a matrix of biodegradable poly(3-hydroxybutyrate) and natural fillers (clay, wood flour, and peat) were constructed [24, 25].

Antifungal activity of the matrix-loaded fungicides and the free fungicides is shown in Figure 10.4. The *in vitro* study showed that all formulations had a significant inhibitory effect on *F. verticillioides*: the size of the fungus colonies was smaller in the presence of the fungicides than in the control group. Photographs show *F. verticillioides* colonies on the nutrient medium in the presence of matrix-loaded fungicides and in the control group. The average radius of the fungus colonies in the Petri dishes with the experimental azoxystrobin formulations was comparable to the radius of the colonies in the dishes with pure fungicide (positive control) (1.3–1.7 cm) while in the dishes without any fungicide (negative control), the size of the fungus colonies was much larger (3.5–3.8 cm). Thus, the experimental formulations were as effective as the free fungicides. Similar results were obtained for the tebuconazole and epoxiconazole formulations. Formulations with different fillers produced similar fungicidal effects. No significant differences in the inhibitory effects on the growth of fungal colonies were found between the pellets and granules.

Biodegradation of the experimental fungicide formulations was studied in laboratory soil microecosystems. Degradation behavior of the experimental formulations and the release of fungicides to soil were studied by determining the mass of the specimens during incubation of formulations in soil, taking into account residual concentrations of polymer and fungicides. These processes are certainly determined by various factors. During the incubation of the experimental granules and pellets in soil,

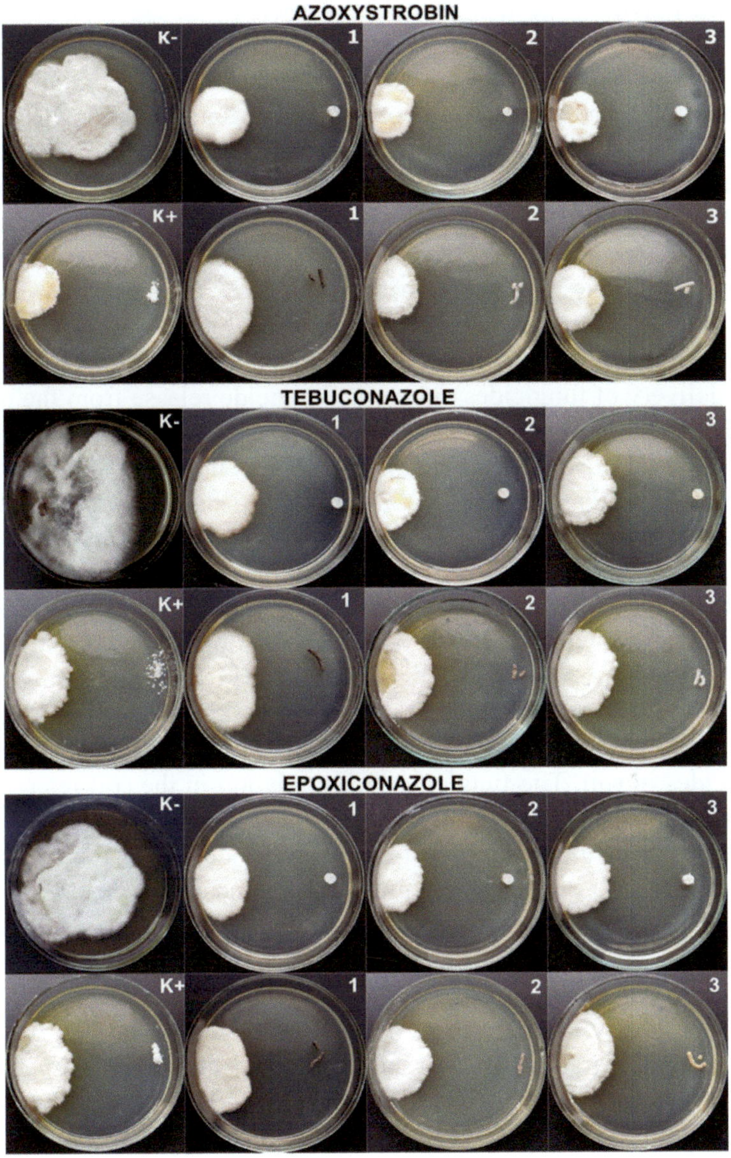

FIGURE 10.4 Sensitivity of *Fusarium verticillioides* to the fungicides in different forms: K–: negative control (without fungicide); K+: positive control (active ingredient), (a) pellets and granules of P(3HB)/peat/fungicide; (b) pellets and granules of P(3HB)/clay/fungicide; (c) pellets and granules of P(3HB)/wood flour/fungicide.

Source: Reprinted with permission from Ref. [23]. © 2019 American Chemical Society.

only the polymer was degraded but the fillers (peat, clay, and wood flour) were not. Even if the formulations were degraded completely, the fillers would stay in the soil. Therefore, a study was performed to investigate the degradation of poly(3-hydroxybutyrate) pellets and granules in soil. The granules were fully degraded in 126 days while degradation of pellets took longer – about 160 days. The type of the filler did not produce any major effect on degradation of the specimens, although specimens containing clay were degraded at a somewhat faster rate. Therefore, the residual mass of clay-containing specimens was the lowest – 55% of their starting mass.

All experimental formulations were capable of functioning in soil longer than the experimental period of 83 days, enabling gradual and slow delivery of fungicides to plants. Fungicide release from the polymer matrix was determined by measuring fungicide contents in the formulations before and after incubation of the specimens in soil. The active ingredients were released rather uniformly, especially from the pellets. By day 20, a significantly higher amount of tebuconazole had been released from the granules than from the pellets. Fungicide concentrations in the experimental formulations with different fillers and fungicide concentrations in the soil were measured during incubation of the specimens in the soil. Fungicide release kinetics was influenced by various factors such as kinetics of degradation of formulations in soil, shape of the specimens, composition of the matrix, solubility of the fungicides, and stability of the fungicides in soil. The fungicides accumulation in the soil released from the granules was somewhat higher than from pellets, especially in the early phase of incubation. The highest fungicide concentrations in the soil with the experimental granules were measured by the third week of the experiment. By contrast, the highest concentrations of the fungicides released from the experimental pellets were observed later, between week five and week seven of the experiment. The dynamics of accumulation of the three relatively poorly soluble fungicides were similar. Measurements of their residual concentrations in the specimens suggested that the fungicides were gradually released from both granules and pellets. After the fungicide concentrations in soil reached their maxima, no dramatic changes in this parameter occurred until the end of the experiment (83 days). The highest concentration of tebuconazole, whose solubility was higher than the solubility of azoxystrobin and epoxiconazole, varied between 20 μg/g and 26 μg/g soil, depending on the filler material (that is 20–26% of the fungicide content in the formulations before soil incubation) and the maximal concentrations of azoxystrobin and epoxiconazole in soil were 15 and 11 μg/g soil, respectively. These concentrations are corresponded to the recommended fungicide rates. The differences in their concentrations in soil were most likely caused by their dissimilar stabilities.

The studies have shown that all fungicide formulations had a pronounced inhibitory effect in vitro against the fungus *Fusarium verticillioides*. The biodegradation behavior of the experimental fungicide formulations in the soil was mainly determined by the shape of the specimens (granules or pellets) without significant influence of the filler type. The content of fungicides present in the soil due to degradation of the formulations and fungicide release was primarily determined by their solubility [23]. The use of P(3HB) and fillers as a matrix for embedding fungicides made it possible to develop slow-release formulations of fungicides, which can function in soil for a long time, enabling gradual and slow delivery of fungicides.

The next study was to develop ecofriendly herbicide formulations. Its main aim was to develop and investigate slow-release formulations of herbicides (metribuzin, tribenuron-methyl, and fenoxaprop-P-ethyl) of different structure, solubility, and specificity, which were loaded into a degradable matrix of poly-3-hydroxybutyrate (P(3HB)) blended with available natural materials (peat, clay, and wood flour) [23]. Studies of biodegradation of the experimental herbicide formulations were performed in laboratory soil microecosystems. The specimens of herbicide loaded polymer/filler carriers were degraded gradually. The residual masses of the pellets and granules at the end of the experiment were 45–70% and 60–80%, respectively, of their starting masses. The composition of the matrix was a minor factor-affecting mass loss of the slow-release herbicide formulations. Degradation kinetics of the majority of the experimental slow-release herbicide formulations was as follows:

$$\ln M_i = \ln M_0 - kt$$

where; M_0 and M_i are the starting mass of the specimen and the mass of the specimen after degradation g; 'k' is the degradation rate constant, and 't' is time, 'h.' The validity of this equation for describing degradation kinetics of all the specimens was supported by the high accuracy of approximation (between 0.91 and 0.98). The highest values of degradation rate constant (0.00746 and 0.0094) were found for the specimens with metribuzin loaded into the clay- and peat-containing matrices and tribenuron loaded into the clay- and peat-containing matrices (0.0074 and 0.0077). Thus, herbicide formulations were capable of functioning in soil longer than the experimental period of 83 days.

The effects of the experimental herbicide formulations on soil microflora were studied. After 83 days of incubation of the slow-release herbicide formulations in the soil, the total abundance of copiotrophic bacteria increased by a factor of 4–10 compared to the control, without herbicide application. The abundance of fungi in soil samples with slow-release metribuzin and tribenuron-methyl formulations increased by a factor of 1.3–3.6 on average while in the samples with the slow-release fenoxaprop-P-ethyl, the abundance of fungi decreased by a factor of 1.2–2.0. The effect of the herbicides on fungi was apparently determined by the type of the herbicide. Examination of the species composition of soil microflora at the end of the experiment (83 d) revealed certain changes in the proportions of the major taxa in soil samples with slow-release herbicide formulations compared to the control soil. *Bacillus* species, constituting up to 28.7%, prevailed in both groups of soil samples. The proportion of *Streptomyces* increased to 21–24.5% on average and *Pseudomonas* to 5.6–12.3%, while *Stenotrophomonas rhizophila* decreased to 8–9.7%. The type of the filler in formulations did not substantially influence the taxonomic composition of the dominant species in microbial communities. Thus, application of slow-release herbicide formulations did not decrease soil capacity for microbial degradation of complex organic compounds, including pollutants.

Herbicides released from the granules reached their peak concentrations in soil after 20 days of incubation while the herbicides released from most of the pellets reached their highest concentrations in soil much later, at Days 48–50. The solubility of the rapidly dissolving metribuzin was 1.2 mg/L, whereas the solubilities of the poorly soluble tribenuron-methyl and fenoxaprop-P-ethyl were 2 mg/L and 0.7 μg/L, respectively. Regardless of the type of the filler, metribuzin released from the pellets reached its peak concentrations in soil after 6–7 weeks of incubation: 60 to 80 mg/g soil, i.e., 50–80% of the loaded active ingredient, and remained steady for a long time, declining insignificantly by the end of the experiment. Residual metribuzin in the formulations was no more than 5–8% of its starting amount, i.e., the loaded herbicide was almost completely released to soil. Concentrations of the metribuzin released from the granules varied in a similar way, but the peak concentration was reached sooner: after three weeks for the specimens containing peat and after four weeks for the granules containing clay and wood flour. Metribuzin release from the pellets corresponded to the Fickian diffusion mechanism and from the granules to the non-Fickian transport, regardless of the filler type, suggesting considerable structural differences between the carriers. Thus, metribuzin was steadily and gradually released from both pellets and granules and, hence, was present in the soil over extended time periods.

So, herbicide formulations enabling slow controlled release of the active ingredients were fabricated. The degree of solubility of the herbicide considerably affected its release rate and degradation of the formulations in soil: the rapidly dissolving metribuzin and tribenuron-methyl were released at faster rates than the less readily soluble fenoxaprop-P-ethyl. Loading of the herbicides into the polymer matrix composed of the slowly degraded P(3HB) and available natural materials enabled both sustained function of the formulations in soil (lasting between one and a half and three or more months) and stable activity of the otherwise rapidly inactivated herbicides such as tribenuron-methyl and fenoxaprop-P-ethyl.

10.4 CONCLUSION

The process of PHA synthesis on glycerol of various degrees of purification has been studied. A technology for the synthesis of PHA by the bacteria *Cupriavidus eutrophus* B-10646 in pilot production has been developed. The conducted studies have shown that the productivity and kinetic characteristics of the strain on glycerol as a C-substrate are comparable with similar processes on glucose. The use of inexpensive and available glycerol as a C-substrate made it possible to significantly reduce the cost of the finished PHA.

In order to further reduce the cost of PHA for agricultural use, mixtures of PHA with various natural fillers have been studied.

Mixtures were formed using P(3HB) and available natural materials (clay, peat, birch sawdust). From mixtures in the form of dry powders and raw paste, mixed forms in the form of granules and extruded 3D forms were obtained; their structure, physicochemical, and mechanical properties were studied. A comparative study of the influence of the type of form and method of manufacture (by pressing powders, extrusion from melts, mortar technologies) on the intensity of biodegradation and the duration of functioning in the soil substantiated the feasibility of using forms that can provide plants with preparations during the growing season (from 1.5 to 3.0 and more months) of two methods – obtaining granules from a crude paste of polymer and filler mixtures and cold pressing of powders. Based on the results obtained and their analysis, a pioneer family of prolonged preparations of herbicidal and fungicidal action for pre-emergence soil application was constructed.

Regardless of the geometry of the developed formulations, the composition of the mixed base, the solubility of pesticides in water, and their stability, it is shown that the achieved concentrations of active substances in the soil are within limits corresponding to the recommended application rates. The results showed the suitability of the developed and studied mixtures and forms of them for the construction of long-term new generation plant protection products.

ACKNOWLEDGMENTS

This study was financially supported by Project "Agropreparations of the new generation: a strategy of construction and realization" (No. 074-02-2018-328) 2018–2019 in accordance with Resolution No 220 of the Government of the Russian Federation of April 9, 2010, "On measures designed to attract leading scientists to the Russian institutions of higher learning."

KEYWORDS

- **bioplastics**
- **biowaste**
- **free fatty acids**
- **glycerol**
- **polyhydroxyalkanes**
- **X-ray diffraction**

REFERENCES

1. Urbanek, A. K., Rymowicz, W., & Mirończuk, A. M., (2018). Degradation of plastics and plastic-degrading bacteria in cold marine habitats. *Appl. Microbiol. Biot., 102,* 7669–7678. https: //doi.org/10.1007/s00253-018-9195-y.
2. Qaiss, Bouhf, R., & Essabir, H., (2015). Characterization and use of coir, almond, apricot, argan, shell, and wood as reinforcement in the polymeric matrix in order to valorize these products. In: Hakeem, K., Jawaid, M., & Alothman O. Y., (eds.), *Agricultural Biomass Based Potential Materials* (pp. 305–339). Cham: Springer, Switzerland.

3. Chen, G. Q., (2010). Plastics from bacteria: Natural functions and applications. *Microbiology Monographs*. Springer-Verlag, Berlin.
4. Chen, G. Q., (2012). New challenges and opportunities for industrial biotechnology. *Microb. Cell Fact., 11*, 111.
5. Volova, T., Shishatskaya, E., & Sinskey, A., (2013a). *Degradable Polymers: Production, Properties, Applications*. Nova Science Pub Inc, New York.
6. Koller, M,, Maršálek, L., De Sousa, D. M. M., & Braunegg, G., (2017). Producing microbial polyhydroxyalkanoate (PHA) biopolyesters in a sustainable manner. *New Biotechnol., 37*, 24–38. https: //doi.org/10.1016/j.nbt.2016.05.001
7. Hermann-Krauss, C., Kolle, M., Muhr, A., Fasl, H., Stelzer, F., & Braunegg, G., (2013). *Archaeal Production of Polyhydroxyalkanoates (PHA) co-and Terpolyesters from Biodiesel Industry-Derived by-Products* (p. 10). Archaea Article-ID 129268. doi: 10.1155/2013/129268.
8. Kourmentza, C., Plácido, J., Venetsaneas, N., Burniol-Figols, A., Varrone, C., Gavala, H. N., & Reis, M. A. M., (2017). Recent advances and challenges towards sustainable polyhydroxyalkanoate (PHA) production. *Bioengineering, 4*, e55. doi: 10.3390/bioengineering4020055.
9. Torres-Giner, S., Montanes, N., Boronat, T., Quiles-Carrillo, L., & Balart, R., (2016). Melt grafting of sepiolite nanoclay onto poly(3-hydroxybutyrate-co-4-hydroxybutyrate) by reactive extrusion with multi-functional epoxy-based styrene-acrylic oligomer. *J. Eur. Polym., 84*, 693–707. https: //doi.org/10.1016/j.eurpolymj.2016.09.057.
10. Gunning, M. A., Geever, L. M., Killion, J. A., Lyons, J. G., & Higginbotha, C. L., (2013). Mechanical and biodegradation performance of short natural fiber polyhydroxy butyrate composites. *Polym. Testing, 32*, 1603–1611. https: //doi.org/10.1016/j.polymertesting.2013.10.011.
11. Angelini, A. S., Cerruti, P., Immirzi, B., Scarinzi, G., & Malinconico, M., (2016). Acid-insoluble lignin and holocellulose from a lignocellulosic biowaste: Bio-fillers in poly(3-hydroxybutyrate), *Eur. Polym. J., 76*, 63–76. https: //doi.org/10.1016/j.eurpolymj.2016.01.024.
12. Khunthongkaew, P., Murugan, P., Sudesh, K., & Iewkittayakorn, J., (2018). Biosynthesis of polyhydroxyalkanoates using *Cupriavidus* necator H16 and its application for particleboard production. *Journal of Polymer Research, 15*, 131. https: //doi.org/10.1007/s10965-018-1521-7.
13. Luigi-Jules, V., Clement, M. C., Alan, W., Des, R., Bronwyn, L., & Steven, P., (2018). Wood-PHA composites: Mapping opportunities. *Polymers, 10*, 751–766. https: //doi.org/10.3390/polym10070751.
14. Du, C., Sabirova, J., Soetaert, W., & Lin, S. K. C., (2012). Polyhydroxyalkanoates production from low-cost sustainable raw materials. *Curr. Chem. Biol., 6*, 14–25.
15. Możejko-Ciesielska, J., & Kiewisz, R., (2016). Bacterial polyhydroxyalkanoates: Still fabulous? *Microbiol. Res., 192*, 271–282.
16. Sabbagh, F., & Muhamad, I. I., (2017). Production of polyhydroxyalkanoates as secondary metabolite with main focus on sustainable energy. *Renew. Sust. Energ. Rev., 72*, 95–104.
17. Fernández-Dacosta, C., Posada, J. A., Kleerebezem, R., Cuellar, M. C., & Ramirez, A., (2015). Microbial community-based polyhydroxyalkanoates (PHAs) production from wastewater: Techno-economic analysis and ex-ante environmental assessment. *Bioresour. Technol., 185*, 368–377. doi: 10.1016/j.biortech.2015.03.025.

18. Green Car Congress (2011). Global biofuels production up 17% in 2010 to hit an all-time high of 105 billion liters. https://www.greencarcongress.com/2011/08/wwi-20110831.html (accessed on 31 August 2022).
19. Statista, (2018). *Leading Biodiesel Producers Worldwide in 2017.* By country (in billion liters). https://www.statista.com/statistics/271472/biodiesel-production-in-selected-countries (accessed on 7 June 2022).
20. Volova, T. G., Shishatskaya, E. I., Prudnikova, S. V., Zhila, N. O., & Boyandin, A. N., (2019a). *New Generation Formulations of Agrochemicals: Current Trends and Future Priorities* (p. 286). Toronto-Canada: CRC/Taylor & Francis: Appl. Acad. Press. ISBN: 9781771887496.
21. Volova, T., Demidenko, A., Kiselev, E., Baranovskii, S., Shishatskaya, E., & Zhila, N., (2019b). Polyhydroxyalkanoate synthesis based on glycerol and implementation of the process under conditions of pilot production. *Applied Microbiology and Biotechnology, 103(*1), 225–237.
22. Volova, T. G., Kiselev, E. G., Zhila, N. O., & Shishatskaaya, E. I., (2019c). Synthesis of PHAs by hydrogen bacteria in a pilot production process. *Biomacromol., 20*, 3261–3270.
23. Volova, T., Prudnikova, S., Boyandin, A., Zhila, N., Kiselev, E., Shumilova, A., Baranovsky, S., et al., (2019d). Constructing slow-release fungicide formulations based on poly-3-hydroxybutyrate and natural materials as a degradable matrix. *Journal of Agricultural and Food Chemistry., 67*, 9220–9231.
24. Kiselev, E. G., Kuzmin, A. P., & Nemtsev, I. V., (2019). Thermal and mechanical studies of biofiller/poly-3-hydroxybutyrate biocomposites. *J. Sib. Fed. Univ. Biol., 12*(3), 302–310.
25. Thomas, S., Baranovsky, S., Vasiliev, A. D., Kiselev, E. G., Kuzmin, A. P., Nemtsev, I., Sukovatyi, A. G., et al., (2020). Thermal, mechanical and biodegradation studies of biofiller based poly-3-hydroxybutyrate biocomposites. *International Journal of Biological Macromolecules, 155*, 1373–1384.
26. Volova, T. G., Zhila, N. O., Shishatskay, E. I., Mironov, P., Vasil'ev, A., Sukovatyi, A., & Sinskey, A., (2013b). The physicochemical properties of polyhydroxyalkanoates with different chemical structures. *Polymer Science, Ser. A, 55*(7), 427–437.
27. Ashby, R. D., Solaiman, D. K. Y., & Foglia, T. A. (2005). Synthesis of short-/medium-chain-length poly(hydroxyl alkanoate) blends by mixed culture fermentation of glycerol. *Biomacromolecules, 6*, 2106–2112. https://doi.org/10.1021/bm058005h.
28. Zhu, C., Nomura, C. T., Perrotta, J. A., Stipanovic, A. J., & Nakas, J. P. (2010). Production and characterization of poly-3-hydroxybutyrate from biodiesel-glycerol by Burkholderia cepacia ATCC 17759. *Biotechnol. Prog., 26*, 424–430. https://doi.org/10.1002/btpr.355.
29. Rodríguez-Contreras, A., Koller, M., de Sousa Dias, M. M., Calafell-Monfort, M., Braunegg, G., Marqués-Calvo, M. S. (2015). Influence of glycerol onpoly(3-hydroxybutyrate) production by Cupriavidus necator andBurkholderias acchari. *Biochem. Eng. J., 94*, 50–57. https://doi.org/10.1016/j.bej.2014.11.007.

Index

A

Abrasive-anticaking agents, 58
Absorption-dominant mechanism, 155
AC conductivity, 100, 130, 133
Accumulated agro-residue waste, 53
Acetabular cups, 18
Acetaminophen, 76
Acetylation, 74
Acid
 alkaline treatment, 49
 base treatment, 30
 environments, 83
Acrylamide (AAm), 79, 82, 181
Acrylonitrile (AN), 75, 79, 82–84, 89, 181
Actuators, 96
Agar medium, 194, 200
Agrowastes, 49, 60
Alicycliphilus sp. BQ1, 86
Alkaline environments, 83
Alkylaminated derivatives, 74
Allylation, 74
Altered gender ratio, 138
Alumina, 18, 123, 153
Ammonium persulfate (APS), 84
Amphiphilic graft copolymer, 87
Anterior cruciate ligaments (ACL), 13
Anticancer agents, 55
Antimicrobial activity, 24, 76
Antistatic agents, 184
Approachable active sites, 54
APTES (3-aminopropyl-triethoxysilane), 51
Articular surfaces, 17
Artificial
 bone graft material, 30
 hearts, 10
 sockets, 21
 wooden legs, 20
Asbestos fibers, 3
Ascorbic acid redox pair, 82
Asymmetric stretching, 30, 205
 vibrations, 30
Atom transfer radical polymerization (ATRP), 87, 178
Attenuation constant, 154
Automobile, 2
Azobisisobutyronitrile (AIBN), 86
Azoxystrobin, 210, 212, 213

B

Bacillus species, 214
Bacterial cellulose (BC), 22, 24
Barium titanate ($BaTiO_2$), 142
Bauxite tailings, 153
Bentonite clay, 51
Benzoyl peroxide (BPO), 88
Betula pendula Roth, 196
Bio-polymer matrix, 1
Biochemical sensors, 58
Biocompatibility, 2, 11, 14, 15, 24, 31, 47, 48, 56, 76, 78
 epoxy resin, 17
 guest molecules, 56
Biodegradability, 1, 3, 10–12, 14, 24, 32, 48, 77, 78, 82, 86, 88, 152, 191, 210
 bio-materials, 14
 materials, 10, 14
 matrix, 3, 10
 polymers, 3
Biodegradation, 210
 rate, 11
Biodiesel, 191, 194
Bioimaging, 51, 58
Biolabeling, 58
Biological
 active molecules, 10
 in vitro tests, 76
 properties, 24
 systems, 135

Biomedical
applications, 8, 51, 58, 78, 106, 177
adhesion, 8
desired properties, 7
self-healing, 8
super hydrophobicity, 8
field, 1, 2, 7, 11, 13, 24
Bioplastics, 216
Biopolymer, 4, 181, 182, 186
functional groups, 181
modification, 177
Bio-resorbable, 11
polymers, 16
BioScadaLab software, 195
Biosensing, 51
Biosignal probing, 48
Biostable biomaterials, 14
Biowaste, 216
Blister disposal, 165
Blood
clotting, 23
coagulation, 8
Bone
fixation, 14, 15
plates, 14
fracture repair, 15
Brunauer-Emmett-Teller (BET), 52, 53
Bulk matrix production, 11

C

Cupriavidus eutrophus B-10646, 191, 194, 197–202, 215
Caffeine, 76
Calcination temperature, 96, 123, 133
Calcium
chloride, 84
deficient hydroxyapatite mineral phase, 30
phosphate glass fibers, 14, 16
Carbon
black, 159
ceramic
composites, 155
epoxy mixture, 14
fiber (CF), 15, 17, 18, 137, 159, 160, 164
polysulfone, 17, 19
reinforced epoxy, 21
woven cloth, 21
fillers, 145
nanofibers (CNF), 145, 152
nanostructures, 151
nanotube (CNT), 137, 145, 156, 158–162, 171
structures, 147
nanowires (CNWs), 156
Carbonate decomposition, 97
Carboxylated, 74, 75
Carboxymethyl
derivatives, 77
reaction conditions, 77
tamarind (CMT), 76–78, 82, 84, 88, 89
graft-polyacrylamide (CMT-g-PAM), 82
kernel polysaccharide (CMTKP), 73, 78, 84, 85
kernel powder-graft-polyacrylonitrile (CMTKP-g-PAN), 84
Carboxymethylation, 74, 77
Carcinogenicity, 15
Cardiovascular
applications, 22
graft, 22
tissue, 22
Casting materials, 15
Catalytically active biomass, 195
Cell
dry mass (CDM), 195
samples dried, 195
Cellulose
aerogel, 150
nanocrystals (CNC), 152
nanopaper (CNP), 152
Cement, 145, 159
ceramic pellets, 163
matrix, 159, 163, 164
Ceramic, 2, 14, 18, 96, 106, 107, 122, 123, 133, 144, 155–158, 170
fibers, 3
matrix composites (CMCs), 2
particles, 112, 113
Ceric ammonium nitrate (CAN), 79, 82, 83, 87–89
Cetyltrimethylammonium bromide (CTAB), 51, 53
Charge aggregation, 98

Chemical
compatibility, 210
composition (bones), 30
free radical initiator, 175, 178, 180, 181
generating free radical sites, 182
induced methods, 175, 178
modification, 74, 78, 175, 176
stability, 18, 155
Chitin
composites, 24
nanocrystals (ChNC), 24
Chitosan, 11
materials, 23
Ciprofloxacin, 76
Cobalt-chrome, 14
Collagen, 15, 22, 23, 30, 155
fibers, 15, 23, 30, 155
reinforced apatite, 30
Commercial
appliances, 135
applications, 144, 186
bulk production, 182
production, 193
Compatibility, 175, 210
Complex
biomechanical movement, 18
impedance
plot, 102
spectrography, 99
spectroscopic study, 95
organic compounds, 214
Composite, 1–3, 5, 7–11, 14–17, 21, 24, 29, 30, 32, 33, 35–39, 41–45, 47, 51, 95, 105, 106, 110–112, 114–117, 121, 135, 137, 141, 144, 146, 147, 150–154, 156, 158, 162, 164, 170, 171, 175, 186, 191, 194
classification of, 2
animal fibers, 3
mineral fibers, 3
natural fiber composites, 3
plant fibers, 4
materials, 2, 14, 15
Compression, 29, 33, 34, 39, 106, 107
Conductive, 5, 84, 95, 96, 100–102, 121, 128, 130–133, 137, 141, 143, 146–148, 151–156, 159–163, 167–170
carbon black (CCB), 146
PPy, 153
Congenital anomalies, 138
Congo red (CR), 87
Contraction, 151
Conventional
free radical graft copolymerization, 181
materials, 165
methods, 55, 147, 192, 195
polymers, 170
solid-state reaction method, 105
Cool dehumidification, 200
Coordination polymerization, 178
Copiotrophic bacteria, 214
Copper bromide-bipyrridine (CuBr-bpy), 87
Co-precipitation method, 157
Corrosiveness, 137, 144
Cosmetic industry, 60
Cristobalite, 48
Critical incorporation concentration (CIC), 160, 161
Cross-linked hydrogel networks, 13
Crystalline, 97, 203, 204, 207, 209
phase, 203, 205, 208
silica nanoparticles, 53
Crystallization
inhibitor, 72
kinetics, 208, 209
temperature, 204
Cu-plating, 167
Cup-type acetabular section, 18
Cyanoethylation, 74, 75
Cyclo-voltammetric sensors, 60
Cytotoxicity, 11, 50, 76

D

Dacron fibers, 11
DC conductivity, 130
Decibels (DB), 143, 144, 146, 148–158, 160, 162–165, 167, 168, 170
Defense technologies, 135
Degradability, 210
Degree of polymerization, 177
Degree of substitution (DS), 75, 77, 83, 87–89
Dendrimers, 48
Detecting agent, 48
Devascularization, 14

Dextrin mediated sol-gel combustion method, 156
D-glucopyranosyl, 71
Diagnosis, 49, 58
Dickson method, 198
Dielectric
constant value, 163
parameters, 121, 133
permittivity, 96
properties, 98, 105, 107, 111, 112, 117, 154, 158, 163
study, 124
Differential
scanning calorimeter (DSC), 74, 76, 79, 83, 85–89, 197, 206, 209
thermogravimetry analysis (DTA), 77, 97
Dimensional stability, 7
Dipolar
polarization, 113
relaxation, 127
Direct
oxidation, 181
sockets, 21
Disc herniation, 16
Dodecyl benzene sulfonic acid (DBSA), 152, 154
Donnan equilibrium theory, 84
Dopamine, 167
Drainage systems, 192
Dried lignocelluloses, 53
Drug
delivery, 47–49, 51, 54–56, 58, 60, 185, 186
carriers, 47, 54, 58
platforms, 51
vehicles, 48, 49
encapsulation, 55
holding capacity, 47
loaded
capacity, 48
hydrogels, 76
Dual cross-linking process, 87
Durability, 14, 20, 21, 146, 165
Dye
ability, 78
binding intermediates, 184
Dynamic
light scattering (DLS), 50, 60, 87, 88
mechanical behaviors, 23

E

E glass fiber, 31–35, 38–41, 45
laminate, 36
Woven Roving WR 400 GSM, 31
Earth magnetic field, 137
Ease of processability, 145
Ecologic orthosis, 24
Economic management, 192
EDX analysis, 42
Effective medium theory, 115
Elastic, 19, 22, 176
modulus, 7, 16, 17, 22
Electric
behavior, 127
conduction, 51, 130, 140–142, 147, 148, 151, 152, 154, 161–163, 165
mechanisms, 156
fields, 137
measurement, 123
percolation, 147
Electrochemical
capacitors, 168
sensing, 48, 49, 54
sensor, 48, 54, 55, 58, 60
Electrochemiluminescence sensors, 48, 54
Electrode polarization, 100, 131
Electroless plating (ELP), 155
Electromagnetic (EM), 135–138, 140–146, 149–157, 159, 161, 164, 165, 168, 170, 171
interference (EMI), 136–138, 140–142, 144, 146, 150, 155, 158, 162, 164, 165, 167, 168, 170, 171
shielding applications, 137, 146, 164, 170
radiation, 136, 137
wave attenuation properties, 170
Electron microscope, 37, 97
Electronic, 2, 136, 151
absorption behavior, 50
band-type conduction, 156
devices, 96, 159

properties, 151
Electrospun
composite fibers, 24
nanocomposite membranes, 24
Elemental analysis, 77, 82, 87, 88
Emboli formation, 23
Energy
dispersive x-ray spectroscopy (EDS), 160
harvesting applications, 116, 117
storage property, 117
Enhanced fluorescence intensity, 58
Environmental
awareness, 176
degradation, 175, 176
friendly methodology, 60
pollutants, 176
stability, 152, 170
unfavorable chemicals, 52
Enzymatic
hydrolysis, 30
oxidation, 75
Epichlorohydrin, 73, 76
Epoxiconazole, 210, 212, 213
Epoxy
composite, 15, 29, 33, 35, 40, 41, 147
Erichrome black T (EBT), 87
Essential amino acids, 70
Esterification, 74
Ethanol, 50, 52, 196, 203
Ethyl
acrylate (EA), 86
methacrylate, 88
Ex vivo bioadhesion time, 84
Experimental
details, 31, 96
fabrication (composite), 31
materials, 31
mechanical testing, 34
sandwich composite fabrication, 33
fungicide formulations, 210, 213
External
fixation, 15
fixators, 15
prosthetics, 20
Extracellular matrix (ECM), 24

F

Fusarium verticillioides, 210, 211, 213
Fabrication
cost, 137
cycle times, 7
Facet degeneration, 16
Fatigue
strength, 18, 29, 45
test, 42
Femoral stems, 18
Fenoxaprop-P-ethyl, 213–215
Ferric
citrate, 201
oxide, 137
Ferrites, 137, 155, 171
Ferroelectric, 95, 96, 99, 101, 102, 105–108, 115–117, 121, 122, 126, 127, 133
hysteresis study, 106
properties, 105–107, 115, 117
transition, 121
Fiber
composites, 3
green composites, 2
reinforced
plastics, 20
sandwich composites, 30
Field emission scanning, 29, 37
electron microscope (FESEM), 29, 36, 39, 87, 166, 167
Fish scale, 29–35, 38–41, 43–45
particulates, 31–35, 39, 41, 43
filled sandwich composites, 35
Flame
resistance, 78
retardancy, 146, 186
Flax fabric, 24
Flexibility, 14, 30, 106, 137, 145, 151, 152, 165
properties, 151
Flexural
energy storing capacity, 21
strength, 16, 29, 41, 43, 45
test, 41
Fluoroquinolone antibiotic, 76
Food applications, 72, 176

Fourier transform infrared (FTIR), 30, 35, 43, 44, 51, 53, 74, 76–79, 82–89, 106, 108, 111, 112, 117, 184, 197, 204
analysis, 43
spectroscopy, 76, 82, 117, 197
Free
fatty acids (FFA), 198, 216
radical
graft copolymerization, 186
polymerization mechanism, 178
polymerization, 175, 178
Freeze-thaw cycling method, 77
Frequency
dependent dielectric constants, 112
independent dispersion, 131
Fungicide, 192, 209–213
concentrations, 212

G

Galactoglucomannan (GGM), 75
Galactomannan (GM), 75
Galactopyranosyl units, 71
Galactosecontaining polysaccharides, 75
Gas
chromatography-mass spectrometry (GC-MS), 75, 196
detectors, 48, 54
Gel
forming ability, 82
mesoporous silica nanodisks, 52
permeation chromatography (GPC), 77, 87
type colloidal suspension, 49
Gelatin, 11
agent, 73
Gene
expression, 48
transport, 48
Genetic materials, 55
Glass
carbon electrode (GCE), 55
transition relaxation, 114, 117
Glutaraldehyde, 73, 76
Glutathione concentration, 56
Glycerin, 197, 200, 202
Glycerol, 76, 191–195, 197–202, 215, 216
Gold nanoparticles (AuNPs), 54
Graft, 13, 17, 69, 78, 79, 82, 83, 87–89, 175, 178–186
copolymer, 78, 82, 84, 89, 175, 177–179, 183–186
product, 184
copolymerization, 78, 79, 82, 83, 87, 88, 175, 178, 180, 181, 185
efficiency, 79, 82, 83
tamarind seed polysaccharide (GTS), 82
Grain resistance, 130
Gram per square meter (GSM), 34, 35, 45
Graphene, 51, 137, 145, 148, 152, 156, 159, 171
foam (GF), 152
oxide (GO), 51, 137, 145, 155, 158, 159, 165, 167, 168
Graphite, 137, 145, 159, 162, 163
Green
composites, 1, 3, 4, 23, 24
house accessories, 209
materials, 176, 186
Guar
galactomannan, 75
gum, 2
Guestholding properties, 55

H

Hardness test, 41
Heat
release rate (HRR), 148
resistance, 176
HEMA (2-hydroxyethyl methacrylate), 23, 88
Hemodialysis application, 13
Herbicides, 192, 213–215
Hexadeca acetyl derivatives, 74
High
density polyethylene (HDPE), 20, 31, 149
energy radiation, 181
technique, 78
flocculation efficacy, 184
frequency (HF), 128, 138, 140
semicircular arc, 100
intrinsic viscosity, 184
performance
anionic flocculant, 82
size-exclusion chromatography-multi-angle laser light scattering, 75

permeable polyester woven-fiber tubes, 11
resolution scanning electron microscope, 108
temperature
calcination, 122
shielding applications, 157
Hip
joint replacement, 14
replacement, 18
Holocellulose, 193
Homogeneous dispersion (carbon fillers), 158
Homopolymer, 181
Honeycomb-distributed WPC ($WPC_{H\text{-}CCB}$), 146
Horticulture, 78
Hosting capabilities, 54
Human
body socket, 21
mesenchymal stem cells (HMSC), 13
Hummers method, 51
Hybrid materials, 175, 176, 186
Hydrogel, 13, 17, 73, 76, 78, 79, 83, 88
morphology, 76
Hydrogen-bonded
interactions, 152
OH group, 205
Hydrolytic degradation, 15
Hydrophilic, 78
carboxamide, 84
Hydrophobic, 15, 18, 50, 56, 155, 205
Hydrothermal synthesis method, 12
Hydroxyapatite, 11, 14, 15, 30, 31
Hydroxyethyl methacrylate (HEMA), 23
Hygienic products, 78
Hysteresis, 95, 97, 99, 101, 102, 108, 115, 116

I

Imaging agent, 48
Imli, 70
Immunosensors, 48, 54
Impedance, 76, 96, 97, 99, 102, 121, 128, 129, 140, 143, 150, 161
analysis, 99
analyzer, 76
spectroscopy, 121, 133
study, 127
Impermeable materials, 11
Implantation, 13
site, 12
Implants, 8
joint replacements, 17
Impregnation cycles, 168, 169
In situ polymerization approach, 154
In vivo drug elution, 56
Incubator shaker, 194, 200
Indomethacin, 56, 76
nanoparticles, 56
Industrial applications, 73, 78
Injection molding, 149
In-situ
chemical oxidative polymerization, 153
nanofiber (ISF), 152
polymerization, 152, 154, 168
Integrated electrical devices, 54
Interconnected filler network, 147
Interfacial
Adhesion bonding, 5
polarization, 113
Interfiber distance, 11
Intermolecular stretching, 205
Internal fixation, 11, 15
Intracellular retention, 56
Intramedullary nail, 15, 16
fixation, 16
Intraocular lenses, 10
Intratumoral diffusion, 56
Intrinsically conducting polymers (ICPs), 151
Ionic
cross-linking, 84
polymerization, 178, 179
IR spectroscopy, 204, 206, 209
Iron nano-sized materials, 137
Irradiation techniques, 175

J

Joint
osteoarthritis, 17
replacements, 10, 17, 20
Jonscher
power law, 102, 131
universal power law, 100

K

Kinetic
characteristics, 215
parameters, 89, 199
Knee replacement, 18
Kyphosis, 16

L

Laboratory
biotechnologies, 193
soil microecosystems, 210, 213
Lab-scale hydraulic hot press, 108
Lactic-co-glycollic acid, 11
Laminate composites, 3
Lead
ceramics, 106, 107
free tungsten bronze niobates, 122
Leather
fibers (LFs), 155
solid waste (LSW), 155
Leg bone fractures, 20
Lichtnecker equation, 115
Lightning activity, 137
Lignocellulosic biowaste, 193
Limit of detection (LOD), 55
Liposomes, 48
Liquid
chromatography-mass spectrometer (LC-MS-MS), 76
crystalline polymer composite, 16
Load-bearing
capabilities, 12
capacity, 16, 23
Low-density-polyethylene (LDPE), 150, 151, 164, 165
Low-energy radiation, 79
Low-temperature cured matrix epoxy resin, 33
Luminescent
materials, 56
sensors, 56
silica nanoparticles (LSNs), 58, 60

M

Magnesium sulfate, 201
Magnetic
materials, 136, 145, 151
resonance imaging (MRI), 17
Magnetite, 137
Materials used, 144
carbonaceous fillers shielding materials, 167
carbon
fillers shielding materials, 146
shielding materials, 151, 155, 160
cement EMI shielding materials, 159
ceramic EMI shielding materials, 155
dielectric filler, 150, 154, 163
insulating polymer EMI shielding materials, 146
intrinsically conducting polymer (ICP)
EMI shielding materials, 151
shielding materials, 168
magnetic filler shielding materials, 149, 152, 156, 161
metallized fabrics, 165
mixed filler shielding materials, 150, 154, 157, 164
textile EMI shielding materials, 165
Matrix
assisted laser desorption-ionization time-of-flight mass spectrometry, 75
loaded fungicides, 210
Mechanical
characterization, 33, 45, 74
degradation, 6
properties, 5, 7, 12–14, 16, 21, 22, 24, 29, 31, 51, 106, 122, 137, 152, 155–157, 160, 161, 165, 170, 184, 215
reinforcement, 147
strength, 73, 150, 165, 176
tester, 76
Medicinal package wastes, 164
Megapascal, 29
Mesochannels, 55
Mesoporous
channels, 54
silica, 48, 49, 55, 56
electrochemical sensors, 49
nanoparticles (MSNs), 48, 51, 54–56, 58, 60
Metabolic
activities, 21
disorders, 17
homeostasis, 138

Metal
chelates, 181
fibers, 3
ion removal, 78
matrix composites (MMCs), 2, 24
nanoparticles, 151
oxides, 150, 151
Methanol, 50, 77, 123, 194, 197
Methyl
acrylate (MA), 87
methacrylate (MMA), 21, 82, 87–89, 181
Methylene blue (MB), 87
Metribuzin, 213–215
Metronidazole, 78
Micro dimensions, 47, 49, 52
Microbial communities, 214
Microstructure investigation, 33
Microwave (MW), 79, 82–85, 87, 88, 138, 140, 156, 160, 161, 181, 182, 185
assisted method, 83, 88, 182
grafting method, 183
induced
method, 182
techniques, 186
irradiated, 87, 88, 182
time, 84
power, 83, 85
technique, 182
Mineral fibers, 3
Mn-Zn ferrite mortar, 161
Mobile
charge carriers, 133, 142
phones, 137, 140
Modern macromolecular chemistry, 78
Modified carbon paste electrode (MCPE), 55
Moldability, 137
Molecularly imprinted polymer (MIP), 55
Monochloroacetic acid, 77
Monomers, 15, 78, 89, 178, 179, 181
Montmorillonite (MMT), 24
Morphotropic phase boundary (MPB), 107, 109
Mouse monocyte-macrophage cells, 50
Mucoadhesive
agent, 74
application, 83
strength, 74
Mucoadhesivity, 76
Multifunctional, 47
graphene composite, 156
Municipal
sewage wastewater, 82
solid wastes (MSW), 149
MXenes, 155

N

N,N-dimethylformamide (DMF), 107
Nanocellulosic substrates, 152
Nanofibers, 24, 152
scaffolds, 11
Nano bimodal particles, 52
Nano-materials, 1
Nanomedicine field, 56
Nanoparticle drug delivery system, 55
Natural
biopolymers, 185
degradable composites, 11
fiber, 1, 3, 24
composites, 3
fish scale particulates, 35
particulates, 45
particulates, 29, 33
polymer, 175
NaturaShield, 165
N-butanol, 50
N-deacetylated chitin, 11
Negative temperature coefficient of resistivity (NTCR), 102, 128, 130
Neointima, 23
Neurobiological mechanisms, 138
Neurological structures, 17
Neuropsychiatric disorders, 136
Next-generation flexible electronic devices, 137
Nickel ferries, 156
Niobates, 96, 133
Niosomes, 48
Nitric acid, 52, 79, 83
Nitrogen
deficiency, 195, 198
free medium, 195, 198

NMR-spectroscopy, 184
Non-biodegradable matrix, 3
Non-corrosiveness, 14
Non-debye type, 100, 130
Non-destructive method, 99
Non-ionic polysaccharide, 69, 71
Nonlinear
  dielectric behavior, 96
  optical behavior, 96
Non-metal
  inorganic materials, 155
  salt, 49
Non-natural bone fabrication, 30
Non-Newtonian pseudoplastic nature, 75
Non-renewable energy resources, 192
Non-resorbable
  composite plates, 15
  materials, 14, 16
  polymer composites, 15
Nontoxicity, 2, 24, 78
Non-uniform dispersion, 154
Novel superabsorbent hydrogel, 84
N-propanol, 50
Nuclear magnetic resonance, 75
N-vinyl-2-pyrrolidone, 83
Nylon
  composites, 21

O

Octadecyltrimethoxysilane (ODTMS), 50
Oil-water separation, 168
One-pot synthesis, 50
Optical
  biomedical imaging, 49
  sensor, 47, 54, 56, 58, 60
Optimum
  attenuation value, 149
  superhydrophobicity, 8
Organic
  acids, 70, 193
  groups, 55
  solvent phase, 51
Orientation, 5, 7
  direction, 7
  nano-dimensional pores, 55
Orthopedic field, 14
Orthorhombic unit cell, 123
Osteoconductivity, 11
Osteosynthesis, 15
Overall rate of polymerization, 89
Oxidation resistance properties, 155

P

Pacemakers, 10
Palm oil fuel ash (POFA), 160
Paracetamol suspension, 88
Parismade bandages, 21
Partially
  cured (PU), 13, 17, 21, 23
  resorbable composites, 16
Particulate
  composites, 3
  filled polymer composites, 35
Peat-containing matrices, 213
Permittivity, 112, 113, 128, 141, 144, 154, 157
Pesticides, 58, 60, 209, 210, 216
Phantom fabric, 165
Pharmaceutical, 78
  applications, 73, 77, 89
  industries, 72
Phase transition temperature, 98, 126
Phenolic compounds, 70
Phosphonate nanoparticles, 56
Photocatalytic
  activity, 58
  stability, 51
Photodynamic therapy, 51
Photo-graft, 88, 89
  copolymerization, 87
Photo-initiator, 87–89
Photo-irradiation, 78
Physical cryogels, 77
Physicochemical
  properties, 24, 201, 208
  techniques, 88
Physicomechanical properties, 14, 24
Physiologic buffer, 76
Piezoelectric, 96, 106
  property, 107
  strain coefficient, 107
Plain cement, 161
Platelet adhesion, 8
Polarization study, 121

Poly (2-hydroxyethyl methacrylate) (PHEMA), 13, 88
Poly d,l-lactic acid (PDLLA), 10
Poly l-lactic acid (PLLA), 10
Poly(3,4-ethylene dioxythiophene) (PEDOT), 151
Poly(3-hydroxybutyrate) (P(3HB)), 197, 201–213, 215
Poly(methylvinyl) silazane, 158
Poly(vinylidene fluoride) (PVDF), 105–117
Poly-3-hydroxybutyrate, 196, 197, 203, 209, 213
Polyacrylamide (PAM), 79, 82, 83, 87, 88, 182
  grafted tamarind seed gum, 88
Polyacrylonitrile (PAN), 84–86, 89
Polyamide, 16
  fibers, 16
Polyaniline (PANI), 54, 151, 153, 154, 170
  strontium ferrite-multiwalled carbon nanotubes (PANI-SrM-MWCNT), 154
Polycaprolactone (PCL), 2
Poly-condensation reactions, 49
Polycrystalline, 122, 124, 127, 133
  ceramic, 95
  crystals impedance spectroscopy, 127
  nature, 124
Polyester
  ether ketone (PEEK), 14–17, 19
  fibers, 15
  knitted fabrics, 21
Polyethylene (PE), 11, 13–15, 20, 23, 24, 31, 115, 117
  glycol (PEG), 11, 23
  terephthalate (PET), 20, 23, 167, 168
Polyglycolic acid (PGA), 2, 11
Polyhydroxyalkanes (PHA), 191–198, 200–203, 206, 215, 216
Polyhydroxyalkanoates (PHAs), 191–193, 201, 205
Polylactic
  acid (PLA), 2, 14, 16
  co-glycolic acid (PLGA), 11
Polymer, 1, 2, 9–11, 14, 16, 19, 47, 54, 77, 78, 106, 135, 137, 144–146, 155, 170, 175, 177–179, 181, 185, 191–193, 201, 202, 204, 207
  blend compatibilizers, 184
  ceramic composite, 106, 116, 117
  composites, 2
  materials, 135, 146, 170, 175
  matrix, 5, 112, 141, 145, 148, 154, 212, 215
    composites (PMCs), 2
  networks, 49
  science, 1
Polymerization techniques, 178
Polymethyl methacrylate (PMMA), 15, 83, 87, 89
Polymorphs, 48, 106
Polypeptides, 24
Polypropylene (PP), 15, 23, 156, 193, 200
Polypyrrole (PPy), 151–153, 168, 169
Polysaccharide, 72, 75, 78, 79, 84, 176, 182, 184, 185
Polystyrene sulfonate (PSS), 151
Polytetrafluoroethylene (PTFE), 18, 23
  fibers, 13
Polythiophene (PT), 151
Polyurethane, 13, 17
Polyvinyl
  acetate (PVAc), 2
  alcohol (PVA), 2, 22, 73, 76, 78, 123, 148, 150, 155
    coated cellulose nanopaper (PVA-CNP), 152
    graphene nanoribbons (PVA-GNR), 148
Porosity, 11, 13, 50, 152, 160, 163
Potassium hydroxide (KOH), 52, 84
Potentiometric sensors, 48, 54, 60
Preterm delivery, 138
Printed circuit boards (PCB), 149
Processability, 145, 151, 152, 210
Prosthetic material, 21
Protein, 2, 47, 55, 70
  malnutrition, 70
Proteolytic degradation, 12
PS sheet function, 149
Pseudomonas, 214
P-type, 48
Pyroelectricity, 96

R

Radiation
induced
graft copolymerization, 180, 181
methods, 175, 178
resistance, 15
Radical polymerization method, 79
Radio
frequency (RF), 136, 138, 171
graphs, 15
Rare-earth
niobates, 122
Reactive chain-end functionality, 179
Recoverable
energy density, 115, 116
storage energy density, 106, 116
Red mud (RM), 153, 154
Redox
polymerization, 178
potential, 56
reagents, 181
Reduced graphene oxide (rGO), 137, 150, 156, 158, 166–169
Relative magnetic permeability, 143
Relaxation frequency, 128
Remnant polarization, 106, 115, 122
Renewable
biological resources, 175, 176
polymers, 78
resource, 4, 176
derived materials, 176
sources, 24, 192
Rice
husk (RH), 49, 52, 53
ash (RHA), 53
plant waste, 47
straw (RS), 49, 52, 53
ash (RSA), 52, 53
Rigakuminiflex, 123
Room temperature hysteresis, 99
Russian Collection of Industrial Microorganisms (RCIM), 194

S

Salicylic acid, 76
Sandwich
epoxy composite, 31, 33
laminated composite, 30
polymer composites, 45
Sawdust particles, 206
Scalar network analyzer (SNA), 144
Scale particulates reinforced epoxy composites, 35, 42
Scanning electron microscope (SEM), 35, 36, 50, 51, 74, 78, 79, 82–89, 95, 97, 98, 102, 108, 112, 113, 117, 121, 123–125, 133, 143, 160, 162, 197, 206, 207
Schlegels mineral medium, 194
Scientific Industrial Testing Research Center, 34
Scoliosis, 16
Second
generation materials, 8
order kinetics, 84
Self-healing
bio-materials, 8
property, 8
Semi-crystalline silica nanoparticles, 53
Shape orientation (reinforcing phase), 5
Shear properties, 13
Shielding
application, 145, 151, 153–155, 165
effectiveness (SET), 142, 143, 156, 157, 168
efficiency, 156, 171
enhancing admixtures, 159
Short length fibers, 7
Sicomin, 24
Silanol groups, 48, 54, 58
Silica, 47–56, 58, 60, 161–163
adsorption properties, 51
core-shell nanoparticles, 54
covering electrochemical sensors, 47
materials, 48, 60
nanocomposites, 54, 58
nanoparticles, 47–51, 53, 55, 58
rich clay, 52
synthesis of, 49
biogenic extraction of silica (rice plant waste), 52
synthesis of silica using tetraethyl orthosilicate (TEOS), 49
Silicon
dioxide (SiO2), 48, 49, 54, 55, 142, 150, 153
rubber, 20

Silk fibers, 12
Silver (Ag), 97, 123, 155, 165, 167
  nanoparticles, 24
Simulated urine (SU), 84
Single-source precursor (SSP), 158, 159
Sintering, 96, 122, 123, 125
Skin
  regeneration, 13, 88
  tissue engineering, 88
Slow-release
  formulations, 213
  herbicide formulations, 213, 214
Smoke
  production ratio (SPR), 146, 148
  suppression, 146
Sodium
  alginate (NaAlg), 87
  hydroxide, 75, 77, 83
  phosphates, 201
  salt (partially carboxymethylated), 83
    tamarind kernel powder
      (Na-PCMTKP), 83–85, 87–89
    TKP containing polyacrylonitrile, 84
Soft tissue replacement, 22
Soil microflora, 214
Solar cells, 58, 168
Sol-gel technique, 49, 50, 53
Solid-state reaction, 96, 105, 133, 157
Solvent casting method, 106, 116, 150
Space charge polarization, 98, 125
Spectroscopic
  analysis technique, 133
  techniques, 50, 83
  tools, 184
Sperm production, 136
Spherical-shaped monodispersed silica
  particles, 50
Spinal
  disorder treatment, 16
  implantation, 16
Spondylolisthesis, 16
Static light scattering (SLS), 75, 77, 82
Stenosis, 16
Stenotrophomonas rhizophila, 214
Stimuli-responsive strategy, 56
Stöber
  method, 50, 51
  protocol, 51
Storage energy density, 106, 115, 117
Strain
  sensing application, 146
  sensitivity, 170
Streptomyces, 214
Stress-strain relationship, 22
Structural
  integrity, 13
  microstructure, 97
S-type, 48
  silica nanoparticles, 48
Succinic, 70
Sugar industries, 193
Sulfonation, 73, 74
Superabsorbent hydrogel, 83–86, 89
Superhydrophobic surfaces, 8
Superior fluorescence lifetime, 51
Surface
  absorbed moisture, 97
  humidifiers, 184
  initiated atom transfer radical polymerization (SI-ATRP), 87
  modification, 186
  morphologies, 29, 45, 50, 96, 98
Swelling properties, 76
Switchable polarization, 96
Symmetric
  bending vibration, 30
  stretching, 205
Synergistic effect
  silica fume, 162
Synthetic
  fiber, 3
  polymers, 175, 185

T

Tailor-made polymers, 8
Tamarind
  fruit, 70
  gum (TG), 70, 76, 77, 87, 95
  kernel
    acetylation, 74
    allylation, 75
    carboxylation sulfonation
      alkylamination, 74
    carboxymethylation, 77

chemical modification, 74
cross-linking (epichlorohydrin), 76
cross-linking (glutaraldehyde), 76
cyanoethylation, 75
degalactosylation, 75
derivatives, 73
physical modification, 77
polymethacrylic acid (TKP-g-PMAAc), 87
polysaccharide, 82, 83
powder (TKP), 69–80, 82, 83, 86, 89
thiolation, 74
mucilage (TAM), 79
products, 70
seed, 70
gum (TSG), 79, 88
polysaccharide, 70, 87
powder (TSP), 70, 71, 74, 76, 82, 83, 87, 89, 146, 148
tree, 69, 70
Tamarindus indica L., 69, 70
seed mucilage (TAM), 79
Tartaric, 70
Taxonomic composition, 214
Tebuconazole, 210, 212
Technological innovation, 138, 176
Telecommunication devices, 137
Temperature
dependent
crystalline forms, 48
dielectric properties, 107
relaxation process, 100
Tendon
ligaments, 23
sheath, 23
Tensile
failure composites, 37
strength, 14, 29, 30, 35, 36, 39, 45, 152, 155, 157, 167
Tension-tension mode, 42
Tetraethyl orthosilicate (TEOS), 47, 49, 50, 52, 55, 60
Tetragonal unit cell, 96
Textile
shields, 165
Theophylline, 76
Thermal
activated
electrical relaxation, 128
mobile charge carriers, 125
analysis, 97
conduction, 155, 156
mechanism, 130
decomposition, 204, 208
temperature, 204
expansion coefficient, 5
insulating materials, 149
properties, 204
stability, 18, 78, 150, 151, 155, 168, 170
Thermogravimetry analysis (TGA), 77, 78, 82, 83, 85–89, 97, 98, 102
Thermoplastic, 15, 31, 192, 204
composite, 15
Thiolation, 74
Thiol-functionalization, 74
Three-dimensional (3D), 11–13, 24, 170, 206, 215
biocompatible scaffolds, 11
cross-linked structure, 177
hydroxyapatite-chitosan nanocomposite rods, 11
Thunderstorm, 137
Tibial components, 18
Tissue
engineering, 10, 12
ingrowths, 12
Titanium, 14, 17, 20, 142
alloys, 14
dioxide, 142
Torsional resistance, 16
Total
cell biomass concentration, 199, 200
heat release (THR), 148
renewable composites, 4
smoke production, 146, 148
Traditional
composites, 1
materials, 9
Transducer, 96
Transesterification, 194
Transfemoral (TF), 21
Transition
metal ions, 181
temperature, 32, 122, 125, 126
Transmission
electron microscopy, 51
power coefficient, 144
Transtibial (TT), 21

Tribenuron, 213–215
methyl, 213–215
Tridymite, 48
Tunable structural properties, 48
Tungsten bronze (TB), 121, 122, 126, 127, 133
crystals, 96
TV broadcasting, 137
Twinflax, 24

U

Ultrafiltration unit, 200
Ultra-high
molecular weight polyethylene (UHMWPE), 13, 18, 20, 23
performance concrete (UHPC), 160, 162
Ultra-sensitivity, 51
Ultrasonication, 50
Ultraviolet
radiations, 88
waves (UV), 50, 89, 138, 181, 182, 184
radiation-induced grafting, 89
Unhydrolyzed grafted products, 82
Uniform
distributed WPC (WPCU-CCB), 146
grain distribution, 102
Urea, 73, 194, 201
formaldehyde resin adhesives, 73

V

Vander Waals forces, 160
Vascular
grafts, 10, 11, 23
prosthesis, 11
Vector network analyzer (VNA), 144
Vietnamese rice husk, 53
Viscosity measurements, 82
Volatile organic compounds, 204

W

Wastewater treatment, 78, 185, 186
Water
insoluble ingredients, 73
repellency, 78
retention, 76
saline retention, 82
solubility, 58
soluble drugs, 76
Wettability properties, 8
Wireless
communication technologies, 135
information transfer, 136
Wood-polyethylene composites (WPCs), 146
Workability, 7
World Health Organization, 70
Wound
dressing, 23
healing agent, 76
Woven roving, 45
WR 400 GSM, 29, 31, 32, 35
Wustite, 137

X

X-ray
diffraction (XRD), 51, 53, 74, 78, 82, 83, 85, 88, 95–98, 102, 106, 108–110, 117, 121, 123, 124, 133, 197, 204, 216
examination, 204, 206
powder diffractometer, 96
Xyloglucan (XG), 73–76, 79, 83
hydrogel film molecular structure, 76
Xylopyranosyl units, 71

Y

Young modulus, 32, 209

Z

Zinc oxide, 142
Zirconia, 18
Zirconium-chitin composites, 11, 12